ETHICAL JUDGMENT: *The Use of Science in Ethics*

Thomas M. Rocco — *1967*

"But after all, what is goodness? Answer me that, Alexey. Goodness is one thing with me and another with a Chinaman, so it's a relative thing. Or isn't it? Is it not relative? A treacherous question! You won't laugh if I tell you it's kept me awake two nights. I only wonder now how people can live and think nothing of it. Vanity!"

FYODOR DOSTOYEVSKY, *The Brothers Karamazov*

ETHICAL

JUDGMENT

The Use of Science in Ethics

By Abraham Edel CITY COLLEGE OF NEW YORK

THE FREE PRESS OF GLENCOE
COLLIER-MACMILLAN LIMITED, London

FIRST FREE PRESS PAPERBACK EDITION 1964

For information, address:
The Free Press of Glencoe
A Division of The Macmillan Company
The Crowell-Collier Publishing Company
60 Fifth Avenue, New York, N.Y., 10011

Collier-Macmillan Canada, Ltd., Toronto, Ontario

DESIGNED BY SIDNEY SOLOMON

To Simon and Fannie Edel

Contents

Preface   *11*

PART ONE—ETHICAL RELATIVITY:
              BACKGROUND AND ANALYSIS

I. *What Is Ethical Relativity?*   *15*

IT ALL DEPENDS . . .—THE THREADS IN THE FABRIC
OF ETHICAL RELATIVITY—INDETERMINACY: THE
BASIC DESIGN.

II. *Barriers and Tools:*
Conscience, Moral Law and Reason   *37*

CONSCIENCE AS POINTER READING—"THE STARRY
HEAVENS ABOVE AND THE MORAL LAW WITHIN"
THE ROMANCE OF REASON—CONTENTS OF REASON
IN ETHICS.

III. *Barriers and Tools:*
The Spectre of the Stubborn Man   *70*

ARE ULTIMATE PREMISES ARBITRARY?—THE "UN-
BRIDGEABLE CHASM" BETWEEN FACT AND VALUE
—ULTIMATE COMMITMENT AND THE CAUSAL
PRINCIPLE—LANGUAGE AS SERVANT, NOT MASTER
—TOWARDS THE SCIENTIFIC STUDY OF ULTIMATE
DISAGREEMENT.

IV. *The Nature and Sources*
of Indeterminacy   92

WHY SEEK TO DIMINISH INDETERMINACY? —
STRUCTURE AND SOURCES OF INDETERMINACY—
TOWARDS COPING WITH INDETERMINACY.

PART TWO—WHAT THE HUMAN SCIENCES CAN OFFER

V. *Biological Perspectives*   115

ATTEMPTS TO DERIVE MORAL PATTERNS FROM
BIOLOGY — THE "NATURALIZATION" OF MORAL
PHENOMENA—THE VALUE OF LIFE—EVALUATION
OF PLEASURE AND PAIN—BASES FOR EVALUATING
BIOLOGICAL DRIVES — EVALUATION OF HEALTH
AND FRUSTRATION—THE FUNCTIONAL HYPOTHESIS
—GROUP ORIENTATION WITHIN THE BIOLOGICAL
PERSPECTIVE — HOW FAR CAN THE BIOLOGICAL
PERSPECTIVE REDUCE INDETERMINACY IN ETHI-
CAL JUDGMENT?

VI. *Psychological Perspectives*   161

CONSCIOUSNESS AND VALUES—THE ROLE OF FUN-
DAMENTAL NEEDS—PSYCHOLOGICAL EVALUATION
OF EGOISM—IMPACT OF A THEORY OF PERSONAL-
ITY DEVELOPMENT—GROWTH OF PSYCHOLOGICAL
CRITERIA FOR NORMATIVE CONCEPTS—PSYCHO-
LOGICAL EVALUATION OF "AMORALITY"—PSYCHO-
LOGICAL SEARCH FOR PHENOMENAL INVARIANTS
—LIMITATIONS IN THE PSYCHOLOGICAL TREAT-
MENT OF ETHICS.

VII. *Cultural and Social Perspectives*   202

CULTURAL AND SOCIAL CONTENT OF MORAL IDEAS
— THE MEANING OF CULTURAL RELATIVITY —

CRITIQUE OF THE THESES OF CULTURAL RELATIV-
ITY—THE SEARCH FOR INVARIANT VALUES—AP-
PLICATION OF BIOLOGICAL AND PSYCHOLOGICAL
CRITERIA—INVARIANT TASKS OF EVERY SOCIETY
—ETHICAL ROLE OF SOCIOLOGICAL DETERMINANTS
AND FUNCTIONS—IS SOCIAL SCIENCE "NEUTRAL"
ON VALUE QUESTIONS?

VIII. *Historical Perspectives*   247

THE TELESCOPIC VIEW — THE STRUGGLE FOR
POWER—THE GROWTH OF MORAL PERFECTION—
LIBERTY AS HISTORICAL ASPIRATION — DEVELOP-
MENT OF MATERIAL CONTROL OVER NATURAL
WORLD — STANDARDS DISCOVERABLE IN HISTO-
RIANS' VALUE CRITERIA—THE MICROSCOPIC VIEW:
THE PROBLEM OF SPECIFIC RELEVANCE — IS
KNOWLEDGE ITSELF CULTURALLY AND HISTORI-
CALLY "RELATIVE"?—APPLICATIONS TO HISTORI-
CAL KNOWLEDGE.

PART THREE—TOWARDS A COMMON ETHIC

IX. *The Theory of the Valuational Base*   293

TAKING STOCK—THE CONCEPT OF THE VALUA-
TIONAL BASE—VALIDATION OF THE VALUATIONAL
BASE—THE VALUATIONAL BASE IN THE EVALUA-
TIVE PROCESS: ILLUSTRATION OF A PERMANENT
INSTRUMENTALITY—THE VALUATIONAL BASE IN
THE EVALUATIVE PROCESS: ILLUSTRATION OF A
COMPLEX IDEAL—THE CONCEPT OF A PERSONAL
BASE—ETHICAL THEORY ITSELF MOORED TO THE
VALUATIONAL BASE—RESIDUAL INDETERMINACY.

Index   341

Preface

The present work is part of a larger study of the foundations of ethical theory on which I have been engaged for some time. One of the aims of the larger study is to restructure the major problems of ethical theory, so as to make clear the scientific and valuational components within the very structure of ethics. This establishes channels of communication with the growing body of the human sciences and the social normative disciplines, thus both enriching ethical theory and increasing its relevance to the contemporary world. This book is only a small part of the general inquiry. It does not present a systematic analysis of ethical concepts; the formulation of a general theory of ethics is the subject-matter of a subsequent work. This book starts within the framework of a particular problem—the highly controversial issue of ethical relativity. Its concern is not to trace the history and relations of that position, but to analyze its general tenet of indeterminacy, and to explore the way in which the human sciences can help provide more definite answers in moral judgment. The result is, in effect, a pilot study in the relations of ethics and the human sciences.

The development of the program of inquiry of which this is one part was greatly aided by a fellowship from the John Simon Guggenheim Memorial Foundation in 1944 and more recently by a grant from the Rockefeller Foundation. I should like to take this opportunity of expressing my appreciation for this assistance.

The fruits of cooperation between ethics and science can come only from large-scale mutual effort. Such efforts have not been frequent, and it seemed worth trying to advance them, whatever the risks and shortcomings unavoidable in attempting to span so many fields. A study such as this can only suggest and sample some of the fruits that a cooperative study might produce. The responsibility for this particular product is, of course, wholly mine, though I owe much of whatever richness it may have to the many colleagues in philosophy and the psychological and social sciences in many centers throughout the country who discussed their work with me in the light of the problems in which I was interested. I should like especially to thank those associated with the Harvard Values Study for giving me access to unpublished manuscripts and research materials.

I am gratefully indebted for critical suggestions to Professors Robert Bierstedt, Horace L. Friess, Yervant H. Krikorian, Helen B. Lewis, Helen M. Lynd, Harry Tarter and Vernon Venable, who read the manuscript in whole or part at various stages; and for very helpful discussion of some of the psychiatric issues to Dr. Joseph S. A. Miller. I am also thankful to Professor Venable for his persuasive urging— after I read a paper on "Scientific Bases for Ethical Judgments" at a Conference at Vassar in 1950—that I make it the subject of this special separate work. I am deeply indebted to my wife, Dr. May Edel, for insight into the problems of the anthropological perspective from the practitioner's point of view, for innumerable special suggestions throughout, and for patient cooperation in reworking the book at every stage of its progress.

Particular acknowledgement is made with thanks for permission to quote passages from several works: to W. W. Norton and Co. for quotations from Otto Fenichel's *The Psychoanalytic Theory of Neurosis;* to Harper and Brothers for quotations from Gardner Murphy's *Personality, A Biosocial Approach to Origins and Structure;* to the New York Academy of Sciences and the authors for quotations from Gregory Bateson and Margaret Mead, *Balinese Character;* and to the editor of *Mind* for quotations from Karl Duncker's "Ethical Relativity?"

A. E.

Ethical Relativity:
Background and Analysis

I. What Is
Ethical Relativity?

§ It All Depends . . .

IS ETHICAL RELATIVITY THE VERDICT OF SCIENCE? The problem of ethical relativity has plagued thoughtful men in all ages. Its special prominence in the modern world stems from the current assumption that it is the last word of science on the question of morality.

Traditional morality rested on supernaturalist or non-naturalist bases. For the most part it was set in a religious framework, yielding an absolute moral order for man. The growth of science and technology in modern times, with their vast accompanying changes in mode of life, ushered in a secular atmosphere. The theoretical pillars of the old moral order were weakened, and serious inroads made upon morality itself. The moral outlook was shaped to the pursuit of this-worldly happiness using the growing instrumentalities made available by scientific discovery. But as the modern world entered upon an era of seemingly endless crisis, more and more voices were heard to sing the praises of a by-gone moral order and to ask, sometimes in plaintive tones, sometimes in wrath, what theoretical foundations science could offer to replace the ones it had brushed aside.

This question met with several kinds of response. Some scientists simply washed their hands of the matter. They declared ethical questions beyond their domain, proper only to religion and philosophy. Some accepted the task

of building ethical foundations. Of these a part regarded
it as an extra-curricular or off-hour assignment, and turned
out hasty hedonisms or grandiose evolutionary schemes
on the basis of a smattering of data from their scientific
work. Others, however, labored intensively to enrich the
tradition of materialist and naturalist ethics which had
come down the ages as the lesser strain in the western
philosophical outlook. But probably the greatest number,
especially in the twentieth century, shrugged their shoul-
ders as if to say that if their work provided no theoretical
foundation for ethics, so much the worse for ethics. One
has to grow up, after all and face the facts of life. Morality
is ultimately arbitrary:

> It all depends on where you are,
> It all depends on when you are,
> It all depends on what you feel,
> It all depends on how you feel.
> It all depends on how you're raised
> It all depends on what is praised,
> What's right today is wrong tomorrow,
> Joy in France, in England sorrow.
> It all depends on point of view,
> Australia or Timbuctoo,
> In Rome do as the Romans do.
> If tastes just happen to agree
> Then you have morality.
> But where there are conflicting trends,
> It all depends, it all depends. . . .

And if these conflicting trends get seriously in each other's
way—well, Hobbes long ago pointed out that where nothing else is turned up, clubs are trumps!

The sharpest test to which such a conception was sub-
jected in our time came with the onslaught of Nazism. In
Nazi theory, to be irrational, to "think with the blood" was
elevated into a professed principle of life. And all that the
defender of reason seemed able to reply was that his own
choice was an arbitrary bias towards rationality. I recall
an able historian in the early nineteen forties arguing that

though he was ready to fight for his democratic values nothing more could ultimately be said for them than for their Nazi opposites — simply that their adherents had been brought up that way!

At this point many philosophers and scientists called a halt. Julian Huxley expressed the insistent feeling:

"We live under the grim material necessity of defeating the Nazi system, ethics and all, but no less under the spiritual and intellectual necessity not merely of feeling and believing but of *knowing* that Nazi ethics are not just different from ours, but wrong and false; or at least less right and less true."[1]

More and more psychologists and social scientists began to look back and protest that they had never really meant what ethical relativity seemed to imply, or else that the sense of ultimate arbitrariness was a product of early scientific results now superseded. Naturalistic and materialistic philosophies began to disown any kinship with ethical relativity. They furnished demonstrations that an empirical approach resting its moral claims on experience rather than on authority and an other-worldly view of man need not accept an imputation of chaos. Even defenders of ethical relativity theories began to look upon the sense of arbitrariness as an interloper rather than a central figure in their theoretical design. But such reconsiderations and recantations served chiefly to fan the fire of a revived moral absolutism which insisted on a theological world-picture as the only safe basis for a human morality which would not revert to brutality.

There is thus a need for a systematic inquiry into ethical relativity. What is the fuller meaning of the doctrine? What scientific facts and theories advanced its career? Is contemporary science committed to it? Is the only alternative a moral absolutism on a nonempirical basis? Or are there a variety of other possibilities? How can its correctness or incorrectness be determined?

---

1. "Evolutionary Ethics," Romanes Lecture 1943, in T. H. Huxley and Julian Huxley, *Touchstone for Ethics* (New York, Harper, 1947), p. 114.

NEED FOR A COMMON-HUMAN MORALITY. To these currents of thought that focus attention on ethical relativity must be added the demand of modern men generally for bases to which they may attach a common-human morality. For problems of war and peace, of national rivalries and group conflicts, of security and change, and hosts of specific issues press upon them no longer as local or provincial questions, but as world-wide challenges. Mankind's very survival depends on solving some of these problems, since conflicts involving them threaten to bring on another world war. Now in untangling the issues men find that there are moral conceptions in the very terms that constitute them. Thus in the admittedly justifiable effort of people everywhere to overcome poverty and disease, there is the ideal of a desirable human well-being. In the upsurge of Asia and Africa there is the ideal of human equality. In facing the issues of capitalism, socialism and communism, there are ideals of human dignity and human freedom, as well as criteria of justice in distribution of material products in a modern technology. If such ideals are to be articulated in a core or minimal common-human morality, are they to be regarded as accidental and arbitrary? Or are they in some sense firmly grounded so that their use in ethical judgment can widen the area of rational decision. And can this ground be found in the study of man as science sees him, or must it depend upon some other theory?

Today the demand for such a theoretical footing is a modest one. It is not a quest for absolutes, not simply a fear of the dark, not a demand for a security that relieves us of decision. It is simply the hope that the great issues of the modern world can be faced with the confidence that there are solutions available to reason and experience. Modern man does not ask for a complete morality. He is used to living with problems, and even to feeling that he may have to struggle against ignorance and intransigence. He knows also that theoretical answers are not easy. But with the tremendous growth and spread of knowledge may he not find some assurance that there are a few firm posts to which a common-human morality can be moored? He does not ask for a detailed map, but he must at least have a compass to guide his ethical course.

§ The Threads in the Fabric of Ethical Relativity

If we are to answer the questions that have been raised we must carry out a full-scale examination of ethical relativity, both as an outlook and as a doctrine. We can best begin by looking more minutely at its texture, by tracing the several threads that run through its fabric, and seeing the general design upon which they converge.

MORALITY A HUMAN PRODUCT. One basic strand in the fabric of ethical relativity is the discovery that morality is a human product. Darwin gave this a solid biological foundation. For if man is part of the animal world, then all his works and all his powers and faculties, however exalted their spiritual quality, must eventually be "naturalized" in the evolutionary process. No matter what ramparts be set up around thought, will, religion, art, morality, social systems, they are ultimately disclosed for what they are and have always been—forms of life and qualities of feeling fashioned by man in his collective effort to make his home on the face of the globe. Ethical relativity is one of the theories of ethics that accepts and even welcomes this discovery. But is it the only one compatible with such a view of man and nature?

The discovery that morality is human is not a new one troubling our day alone. There are significant echoes and re-echoes of this discovery and its implications back in ancient Greek times. There is, for example, a famous passage in the Greek historian Herodotus, in which he appraises the relativity of custom:

"Darius, after he had got the kingdom, called into his presence certain Greeks who were at hand, and asked 'What he should pay them to eat the bodies of their fathers when they died?' To which they answered, that there was no sum that would tempt them to do such a thing. He then sent for certain Indians, of the race called Callatians, men who eat their fathers, and asked them, while the Greeks stood by, and knew by the help of an interpreter all that was said—'What he should give them to burn the bodies of their fathers at their decease?' The Indians exclaimed aloud, and bade him forbear such language. Such is men's wont herein; and Pindar was right, in my judgment, when he said, 'Law is king o'er all.' "[2]

Herodotus is expressing the belief that the Persian king Cambyses was mad for mocking men's customs, since every people sets a high store on its ways. On the other hand, in the historian Thucydides, we find the denial that there is an eternal justice or intrinsic right attacked as a doctrine that might is right.[3]

The Greek dramatists also face the issue of the loss of ultimate standards. The comic poet Aristophanes, for example in his *Frogs*, caricatures the tragedian Euripides. He is presented as a corrupter of morals, a self-seeking demagogue who is ready to break an oath with the flippant remark that "Only my tongue promised." In one passage, the god Bacchus expresses his ardor for Euripides' work to the hero Hercules:

"*Hercules:* Do you like that kind of stuff?
Bacchus: I'm crazy after it.
Hercules: Why, sure it's trash and rubbish—Don't you think so?
Bacchus: 'Men's fancies are their own—Let mine alone'—
Hercules: But, in fact, it seems to me quite bad—rank nonsense.
Bacchus: You'll tell me next what I ought to like for supper."[4]

---

2. The History of Herodotus, Bk. III, 38, translated by George Rawlinson (New York, Everyman's Library), vol. I, pp. 229-30.

3. Especially in his account of the dialogue between the Athenian ambassadors and the Melian leaders (*History of the Peloponesian War*, Bk. V, ch. 17).

4. Translated by J. H. Frere, *Four Famous Greek Plays* (New York, Modern Library), p. 218.

Note the assumption of the relativity of taste, and the ready resort to food as an analogy. It is very much like recent ethical writings in which a man's morals are his own, like his choice of port over sherry. The best comment comes from the slave Xanthias who is standing by and says: "But nobody thinks of me here, with the bundles."

Greek philosophy tended to line up in opposing camps on the question of ethical relativity. Many of the Sophists held that man is the measure of all things, that his morality as well as his language, science and manners, is manmade. It has no ultra-human reference. Different people have their different ways. There is nothing "natural" about it. Not necessity but accident plays a part in determining the ways of man. On the other hand, for Plato, there is a natural order for men, an absolute justice and an absolute good which the true philosopher can discern. Relativity characterizes moral ignorance, not moral truth. For Aristotle, too, there is a nature to things, and this means a natural justice. His applications of this have a curious flavor today, reenforcing the relativity against which they were aimed. For to Aristotle slavery was part of the very order of nature, while money-lending was contrary to it. Apparently there is more to the dispute of custom or convention versus nature than a conflict of pure categories.

Another approach is to be found in Sextus Empiricus, the sceptic of the second century A.D. The natural in ethics is rejected on pragmatic grounds. If you believe in a natural good and evil you will be committed to a strenuous life of pursuit and avoidance, and therefore condemned to unhappiness.

"But if a man should declare that nothing is by nature an object of desire any more than of avoidance, nor of avoidance more than of desire, each thing which occurs being relative, and, owing to differences of times and circumstances, being at one time desirable, at another to be avoided, he will live happily and unperturbed, being neither exalted at good, nor depressed at evil, manfully accepting what befalls him of necessity, and being liberated from the distress due to the belief that something evil or good is present."[5]

Therefore, the advice is, one conducts one's life empirically and undogmatically in accordance with rules and beliefs commonly accepted, an outcome resembling Herodotus' respect for custom. Sextus, with all his emphasis on the relativity of morals, does not suggest that some people may simply prefer the strenuous life.

In many respects the rediscovery that morality is a human product cut even more deeply into ethical thinking in modern times than did its first establishment in the ancient world. The western world had made a tremendous investment in the separate identity of spirit. Its religion, its modes of justifying its institutions to itself, its thought about human relations, had all enshrined the supernatural bases of its moral outlook. To remove this seemed to take away the keystone and bring down the whole edifice. The fact that it did not fall was often felt as a constant moment-to-moment postponement of impending doom.

EVERYTHING CHANGES. That morality is a human product is one strand in ethical relativity. Another basic strand in the pattern is the expectation of change.

The ancients had strong feelings about change, but to them it was an enemy. It was not the real core of the world. "There is nothing new under the sun," says Ecclesiastes in the Old Testament, and change to him means age and death. In the fall from Paradise, change is deterioration from man's fixed good. Aristotle looks longingly at the perfection of the heavens in contrast with the disorderly movements of the sub-lunar realms. In any case, behind all change there is a plan—an eternal changeless nature for the universe and in one fashion or another the fixed moral law for man. This was the tradition that dominated the ancient and the medieval world. Even in the 18th century, when change found expression in the ideal of prog-

---

5. *Sextus Empiricus*, translated by R. C. Bury, vol. III (New York, Putnam's, Loeb Classical Library, 1936), p. 443.

ress, men still thought in terms of endless approach to a
definite goal for mankind.

Modern philosophies since the rise of evolutionary
theory in the 19th century have come to give change a
more central place. They see constant change in the cosmos, in the animal world, in populations and social forms.
Even the most abstract philosophies have begun to think
more in terms of process and time and the flow of events,
and less in terms of a fixed essence and a rational nature
to the world and man. Modern practical attitudes reflect
the actual change that a constantly revolutionized technology brings in human living. Unlike the medieval cathe-
dral, built to last beyond the memories of men, a modern
skyscraper is built to be replaced in due time. A home is
not an ancestral dwelling, but a rapid construction that
may not survive the last mortgage payment—if you have
not moved away by then. The number who live today in
the house in which they were born is small, the number
who glory in the fact is smaller still. We streamline everything and turn to the latest model. Why make an exception
of morality? Is not the quest for fixity a snare and a
delusion?

INDIVIDUALISTIC EGOISM. A third strand in the pattern
of ethical relativity is individualistic egoism. "It's every man
for himself and the devil take the hindmost." (But since
nobody looks behind at the other man, Satan picks up
quite a few extra!) Not all ethical relativity is self-centered.
It may be culture-centered or nation-centered. But much
of it is individualistic, and here it is strengthened by the
egoistic theory that one's own welfare or happiness or
pleasure is one's ultimate and inescapable standard for
judging good and evil, right and wrong. Historically, such
a view has been associated with the rise of acquisitive and
competitive patterns in a commercial economy. In the
modern world after the break-up of feudalism, the acquisitive passions achieved first respectability and then pre-
eminence. R. H. Tawney has traced, in his classic *Religion
and the Rise of Capitalism*, the shift from the system of

ordered duties and relations in the hierarchical Catholic morality of feudalism to the virtues of thrift, diligence, sobriety, frugality in the Calvinist scheme, and then to the break-away of individual acquisitive patterns from all religious control.

Hobbes' *Leviathan* in the 17th century represents the first egoistic ethics that is systematic, outright and secular in its egoism—egoist through and through. His ethical theory is egoist in form: it defines "good" in terms ultimately of individual appetite. It is egoist in content: men pursue gain, safety and reputation. And it is egoist in the last details of psychology: fellow-feeling is identified with pity, and this stems from the imagination that a like calamity may befall oneself! More than a century later Bentham fashioned this egoism into a full-fledged social philosophy. In Bentham's hedonism, a man inevitably pursues his own pleasure and avoids his own pain. These are his profit and loss. The social standard—the greatest happiness of the greatest number—emerges from each man's pursuit of his own gain. Artificial legal sanctions add to natural identity of interest. Virtues are safe investments which, like honesty, prove to be the best policy. A democracy of acquisitive individuals is the outcome. How philosophers sought to soften this harsh egoism, to override or refute it, is another story. But its central individualism—each man the ultimate judge in his own case—is woven into the fabric of ethical relativity.

THE STRUGGLE FOR POWER. Not always distinguishable from this last thread in the pattern of ethical relativity—at least running parallel with it all the way—is the emphasis on the struggle for power in political theory from Machiavelli to the present day. Machiavelli made the pursuit of power a distinct end, irrespective of moral judgments of means. But he really thought beyond power, with an eye on the fruits that power would bring for the Italian states of his day. Similarly, Hobbes was careful to point out that man's perpetual and restless desire for power rested on his insecurity in maintaining what he has. But modern Machia-

vellian theory has shown no similar restraint. Power is made
the end in itself, and used as a key to explain the conflicts
of men. What is more, ideals, social programs and even
philosophies are seen as competing symbols to win over
masses of men to support one's domination. Marxian historical investigation has attempted to show that the dominant morality of a society reflects the needs of its dominant
class, but Machiavellian power theory is not bound to the
avowed Marxian aim to abolish exploitation. It regards exploitation as inevitable, and the only question is who shall
be the exploiter, who the exploited. With apparent neutrality a Pareto or a Mosca will describe the techniques of
a ruling class in holding to power, and Mosca in his book
The Ruling Class, even defines a democracy as a ruling
class that keeps itself open so that individuals may rise into
it from below. While power theory of this sort is not a part
of all relativistic outlooks, it does add its underlying thread
to the pattern.

MECHANISTIC PSYCHOLOGICAL AND EDUCATIONAL THE-
ORY. A fifth strand of thought without which ethical relativity would never have reached its present proportions, was
contributed by mechanistic psychological and educational
theory. Specifically, from John Locke in the 17th century
to the behaviorism of J. B. Watson in the 20th century,
there is the wide-open view that man, within the limits of
his physical powers, can be fashioned in unlimited ways.
Locke attacks any inborn or innate moral principles in the
human mind. The mind is like a wax tablet, unmarked to
begin with. Experience and experience alone fashions the
pattern of grooves. As proof he culls conflicting customs
from the anthropology of his day, and conflicting theories
from the philosophers. Behaviorism takes the same position
in physical rather than mentalistic terms. As Pavlov conditioned his dogs in the laboratory to salivate at the ringing
of a bell instead of the presence of food, or at the sight of
a spot of light, so too men may be conditioned from their
birth up by the careful play of stimuli until they assume
any planned form. Men's beliefs, men's attitudes, and men's

values therefore reflect the conditioning milieu. Educational theory thinking in this vein casts its concepts in terms of indoctrination, propaganda and counterpropaganda, not in terms of developing a natural moral order or bringing to fruition inherent human capacities.

CULTURAL RELATIVITY. Cultural relativity is the sociological and anthropological contribution to the pattern of ethical relativity. The lesson of Herodotus, of Sextus Empiricus, of Locke, can be read in the blunt statement of Sumner's *Folkways*, the classic of American sociology: "It is most important to notice that, for the people of a time and place, their own mores are always good, or rather that for them there can be no question of the goodness or badness of their mores. The reason is because the standards of good and right are in the mores."[6] And he entitles one chapter "The Mores Can Make Anything Right and Prevent the Condemnation of Anything." More recently, Ruth Benedict's *Patterns of Culture,* presenting the variety of systematic patterns into which the range of human needs and feelings have been fashioned in different societies, concludes: "We shall arrive then at a more realistic social faith, accepting as grounds of hope and as new bases for tolerance the coexisting and equally valid patterns of life which mankind has created for itself from the raw materials of existence."

LINGUISTIC INFLUENCES. In the last few decades, a linguistic thread began to weave its way into the design and it has reached surprisingly large proportions. Semantic studies asked for the *meaning* of judgments of good and evil, right and wrong. Are they scientific statements that can be verified by logic and observation? Or are they emotional expressions to which one cannot significantly attribute truth or falsity? Thus A. J. Ayer, in his influential book, *Language, Truth and Logic,*[7] maintained that "Stealing is wrong" is no more than an expression of horror attached

---

6. W. G. Sumner, *Folkways* (Boston, etc., Ginn, 1934), I, 65.
7. London, Gollancz, 1936, ch. VI. See also comments in preface, 1946 edition.

to the concept of stealing, as if one said "Stealing!!" C. L. Stevenson's *Ethics and Language* presents the thesis that so-called ethical judgments are partly descriptive of one's personal attitudes, partly attempts to influence others; for example, "This is wrong" is most serviceably interpreted as *I disapprove of this; do so as well.*[8] The august imperative quality of the traditional moral law is enfeebled in this picture of personal fiats attempting to seduce one another.

THE CONVERGENT EFFECT. This is a brief summary of the strands in the blurred pattern that emerges in the fabric of ethical relativity as it has grown up today. Some of the strands are themselves confused and even at times contradictory—for how can there be a natural power drive if man is purely plastic? If we are to analyze the pattern, to understand and assess it, perhaps to correct for it, we need to evaluate the separate strands. It will be part of our task in this book to see what the current results of the sciences do to these separate strands—how, for example, the psychologist views egoism or the power drive, and whether history sees any constancies in the pattern of human striving. In the course of this analysis, some of the strands may be pulled out and discarded, others rewoven. But beyond analysis of the strands, we must examine the underlying texture and design of the whole fabric, and see the doctrine in which it issues.

There can be no doubt as to the convergent effect. The design that is formed is one of arbitrariness in ethics. This is the focus of much of popular and philosophical thinking on the question. For human values, human goals, human aspirations come to appear as products of self-seeking, of cultural indoctrination, of power fights, of man's "mere" animal nature. If morality is man-made and if it is changing, there is no purpose metaphysically or theologically imposed. If egoism is correct, there are no morally mandatory social obligations not subject to individual veto. If men are constantly pursuing power, morality is a pawn in the

---

8. C. L. Stevenson, *Ethics and Language* (New Haven, Yale University Press, 1944), p. 21. For analysis of this view, see below, pp. 84 ff.

general strategy. If the mind is a clean slate or the body an almost wholly plastic material, then almost anything is possible. If cultural relativity is a sociological truth, then your morality is a function of your domicile. If moral assertions are simply expressive, it all depends on what you feel. In any case, it all depends. . . .

This arbitrariness, however, does not always yield the same consequences for human attitude and human action. It may breed cynicism or acceptance in modern times, as it did for the ancients. For what if it does all depend? What "and therefore . . ." follows? Morality is man-made; and therefore you do not have to fear the gods? Or, and therefore you need not take it so seriously? Or, and therefore man can change it? Morality is changing; and therefore there is no use holding on? Or, and therefore look to what is coming? Or, and therefore jump into the driver's seat and steer? Morality expresses self-seeking; and therefore you may give free rein to your desires? Morality is ultimately a struggle for power; and therefore don't be fooled by pious sentiments? Or look out for yourself? Any morality can be written on a clean slate; and therefore man can remake himself by changing conditions? Or man can acquire power over others by appropriate indoctrination? Morality is relative to culture; and therefore, when in Rome do as the Romans do? Or don't feel bound by an abhorrent custom? Or do what you feel like but conform outwardly? Moral assertions are simply expressive; and therefore it's all up to you as an individual? Or we must respect one another's feelings? Or morality is just froth?

It seems that the pattern in the fabric has quite different moral and social potentials, that it too is relative and depends. . . .

§ Indeterminacy: the Basic Design

WHAT IS THE ESSENCE OF THE RELATIVIST DOCTRINE? It is not easy to state the essence of ethical relativity as a doctrine, even though its components and the bases upon which it rests are clear. Neither its adherents nor its opponents have really pin-pointed a limited formulation of the issue upon which there is wide agreement.

Is the essence of ethical relativity to be found in some dispute about the "reality" of moral values, as it is so often posed? Does it lie in the admitted variety of moral outlooks, in their conflict, in some uniqueness of each particular morality, in the "subjectivity" of moral experience? Or is it in something else to which all these point?

Much of the conflict of absolutist and relativist when they match doctrines in discussing moral problems, moral answers, and ethical theories, does seem to focus on concepts of the real and the genuine. One speaks of genuine moral issues, the other simply of people being troubled. One pictures an absolute good as the ultimate object of human aspiration which alone will satisfy man's striving; the other says we are dealing merely with the paths of men's changing preferences. Yet when we examine such formulations carefully, the dispute seems to be less about the reality of specific moral phenomena than about their interpretation. Do they point to an eternal order or to a changing scene? Are they unified or hopelessly diversified? Is there a fixed human nature? Are the phenomena intrinsically individual or capable of common establishment? So that it is here that the essence of the doctrine of ethical relativity is to be sought.

Is the essence then variety? It cannot be the fact of variety alone, although this emphasis is most common in general discussions. For even if we grant the greatest variety—in what are felt to be moral problems, given as moral answers, offered as ethical theories—this variety cannot simply be equated with ethical relativity. For there are certainly many areas in our world where variety occurs without being construed as a kind of relativity. No doctrine of botanical or zoological relativity is proclaimed because there are different species of plants or animals.

Nor is the fact of conflict the heart of the doctrine. For conflict may be pervasive and yet there be some way of determining which side is right. And uniqueness is not enough, for wider systematic formulations may be possible, so that every instance becomes a particularization of some general law. Subjectivity, too, will not suffice. For whatever precisely "subjectivity" may mean, sense experience has it as well as moral experience, and this does not forestall an imposing structure of orderly science.

The essence of the position lies beyond this descriptive level of variety, of uniqueness, of conflict, of subjectivity. It lies in the assertion that these are in some sense ultimate, irreducible, irreconcilable; that there is no pattern or generalization of which the varied and the unique are but different expressions; that the conflict can never be resolved; that there is no moral bridge to break the individual's isolation. In short, that there are no common moral questions, and above all, no common moral answers.

And this appears to be the essence of the problem which modern man and modern ethics must face. Are there any answers? Does the universe as science pictures it really leave us stranded with all our questions and all our answers indeterminate? Or are there some answers, some guideposts or moorings for morals? Indeterminacy—the fact of no definite answers available or achievable—seems to be the heart of the relativist position in ethical theory.[9]

A PECKING ORDER AMONG ETHICAL THEORIES. On this formulation of our central problem, relativism and absolutism are seen to take their customary places at opposite poles. One asserts that all is indeterminate; there are no answers nor any way of getting at them. The other asserts

9. For the reader who may associate the term "relativity" with the physical theory that goes by that name, it is important to point out that the ethical use has been a quite distinct one. In the history of human thought about morals, ethical relativity is a phenomenon whose very core seems to be the belief in unavoidable indeterminacy. We therefore have little choice but to call it by the name it has had, and warn against confusing it with something else. The physical theory of relativity serves, on the contrary, as a way of introducing greater determinateness. Only in popular misinterpretations does the term suggest subjectivity and arbitrariness. This point is clearly brought out in Philipp Frank's little book *Relativity, A Richer Truth* (Boston, Beacon Press, 1950).

that there are answers, written in men's conscience, or the dictates of the gods—somehow available for his grasping. The variety and uniqueness are surface phenomena or false ones. Along this range are spread the familiar theories of the ethical tradition. They fall into a kind of pecking order from the most absolute absolutism to the most relative relativism, and each can point the finger of relativistic scorn at the one below it. In the upper regions indeterminacy is cast out of doors, or at least hidden in the cellar. At the bottom, indeterminacy is in the very hearth of the dwelling. It is almost as if it had been put into the foundations with a sign on the doorway: Expect an earthquake momentarily, all ye who enter here. But has even the most absolute of absolutisms exorcised possible indeterminacy? And has even the most relative of relativisms guaranteed indeterminacy a permanent residence?

If we are to seek clues to the nature of ethical indeterminacy in the way ethical theories have handled it, we must look not merely at the general propositions which they put in the show-cases, but at the modes of knowing, interpreting and applying they provide. It is the better part of wisdom not to buy a complex contraption without making sure there are instructions for use. Otherwise we might find an ethical theory advertising itself as absolutely without indeterminacy, but in fact it might consist simply in the proposition that "Sin is evil," without telling us what a sin is or how to recognize evil.

Locus of Indeterminacy in Absolutist Theories. Some absolutistic theories speak as if we can simply intuit what is good and what is right by exercising our human sensitivity. Others take a religious form and have moral answers depend upon the nature of God. Still others, such as the Kantian ethics, have morality issue in absolute imperatives which are universal laws. But none really escape all indeterminacy.

The intuitionist theories even while they make values *ab-solute* in the literal sense of cut off or released from any kind of dependence whatsoever, are extraordinarily lax in their interpretation of intuition. Thus if you have a different intuition from me, I can only try to help you see more clearly, or dismiss you as morally insensitive.[10] Hence by removing all indeterminacy from the values such absolute intuitionists have simply transferred it to the mode of cognition.

The religious theories have their dependency-points as well. Morality *depends* upon God, which is in the first place, abstractly speaking, a kind of relativity. This tends to be overlooked largely because God is pictured as so dependable, and morality is seen as issuing from His rational nature. Once in a while, a Duns Scotus argues that God does not will what should be because it is right, but it is right because He wills it. This makes His will primary and possibly arbitrary. But this indeterminacy is overcome by revelation by which the path of will becomes dependably known. A truly unknown and indeterminate source would leave men in the kind of dread and perpetual anguish that contemporary existentialists love to parade. For if one had to gamble on a perpetually and unpredictably changing divine will about right and wrong, the least act would bear the chance of sin. Revelation, however, only postpones the indeterminacy. It is transferred to the diversity of God's preachers; for different religions have different revelations, and even the same account allows of different interpretations. The indeterminacy here stems not from the mere fact of disagreement, but from the fact that there is no definite mode of decision by which, even if only in principle, one could settle the issue.

A morality of absolute laws gives the appearance of completely removing indeterminacy. But it too simply covers it over. There are indeterminate elements in the way the particular laws are decided on. There are further ones in the meaning of the terms in the laws. The more you insist that it is always wrong to tell a lie, the more prone you will be

---

10. Examples of these theories are to be found in G. E. Moore's *Principia Ethica* (Cambridge University Press, 1903), pp. 143-146, and N. Hartmann's *Ethics* (translated by Stanton Coit, New York, Macmillan, 1932), I, p. 39. Cf. below, pp. 189 ff.

to render the definition of "lie" indeterminate.[11] Thus one finds the argument that where it is justified, an apparent lie is not really a lie but sparing another's feelings, or exercising mercy, or whatnot. And if one avoids this type of vagueness, the indeterminacy reappears in the conflict of laws applicable to the same situation. If saving a life is right and lying is wrong, what should be done where a lie is the only way to save a life? This whole question of the nature of moral laws will pursue, if not haunt us in what follows.

LOCUS OF INDETERMINACY IN EMPIRICAL THEORIES. Ethical theories that appeal to experience instead of to intuition, religion or reason, sometimes treat indeterminacy simply as a problem of difficulty in measurement. They may call attention to differences in what is right or good under different psychological, historical and individual conditions. In that sense morality is relative.[12] A certain action is wrong when done by a normal man, but not if done by an idiot who cannot make fine distinctions. You cannot expect the same rules of private property concerning land in a primitive hunting society and a small individual peasant society. You cannot impose the obligation to save a drowning man by swimming out for him upon one who cannot swim or who is desperately ill. Such individualization and particularization hardly constitute indeterminacy. For they call attention to the factors which make single and decisive answers correct; in short, they remove rather than enshrine indeterminacy.

The same is true of Utilitarian ethics, at least in the 19th century, when it judges what is right by what yields the greatest happiness of the greatest number. It too assumes that there is in every case a correct answer which skillful moral arithmetic could calculate. The father of Utilitarianism, Jeremy Bentham, was quite vehement in his attacks on previous standards of morals as arbitrary expres-

---

11. This shift in the locus of indeterminacy is well analyzed in Philipp Frank's *Relativity, A Richer Truth* (Boston, Beacon Press, 1950), p. 47.

12. See, for example, Harold Höffding, "The Law of Relativity in Ethics," *International Journal of Ethics*, I, 30-62 (Oct., 1890).

sions of sympathy and antipathy, and strange as it may sound to modern ears (for which the appeal to pleasure is an appeal to caprice) he regarded maximum pleasure measurements as completely objective.[13] Indeterminacy nevertheless reared its head at a number of points. For one thing, the Utilitarian slogan depends upon a democratic equalitarianism; it simply assumes that every man counts as only one in pleasure computations and equal to every other man. In the second place, indeterminacy appears in going from social judgments to individual cases. Granted honesty is generally the best policy, is it the best policy for a particular man in a strategic position? Appeals to natural identity of interest or resorts to sanctions to make interests compatible may describe mechanisms of possible adjustment, but they do not remove the indeterminacy of individual judgment if every man inevitably pursues *his own* pleasure.

INDETERMINACY ENSHRINED IN SUBJECTIVIST AND EMOTIONAL THEORIES. In the theories that boast of their relativity, the danger is not that indeterminacy may be conjured away a priori or effectively disguised. On the contrary, we have to look out for a too ready assumption a priori that indeterminacy is inevitable. The various subjectivistic and naturalistic theories that tie morality to men's emotions, feelings of approval, preferences, are prone to postulate the arbitrariness of the emotions. It was not always so. Earlier moralists—such as Adam Smith in the 18th century—assumed a uniform structure to men's ultimate moral emotional reactions. But the study of cultural and individual variation in emotional response, as well as in the objects to which emotions are attached, gave a strong relativist turn to theories cast in terms of the emotions.[14] However, this ultimate indeterminacy need not follow. It depends on how the variability is to be understood, whether inconstancy or indefiniteness is inherent in the nature of the subjective and the

13. Jeremy Bentham, *An Introduction to the Principles of Morals and Legislation* chs. 2, 4.

14. Cf. Edward Westermarck, *Ethical Relativity* (New York, Harcourt Brace, 1932), pp. 216-217.

emotions. Even in the absence of gross uniformity, individualization and particularization may rest on determinate grounds. There is some tendency today to challenge the established treatment of emotions as disorganized response and to suggest that they constitute a kind of motivation process arousing, sustaining and directing activity.[15] This would open the way for a fuller study of their structure and "logic." Certainly the development of depth psychology has shown that even the surface illogicality of the emotional life may be understood in terms of a fuller dynamic psychological-biological picture. It follows that a view of ethics giving a primary place to the emotions cannot now be sure in advance whether determinacy or indeterminacy will be the outcome. For it all depends on the real character of human emotions, and this still requires fundamental psychological study.

There is one way to ensure an a priori triumph for indeterminacy and that is so to reconstruct ethical language that there will be no ethical judgments! This is done in extreme forms of contemporary emotive theory. Ethical judgments are seen as disguised *expressions* rather than propositions. If by saying that something is wrong you are just expressing your horror, you are not saying anything factual, hence there is no problem of truth or falsity, no question whether value judgments admit of scientific verification.[16] We shall have to examine later on this attempt to enshrine indeterminacy by use of one form of language in ethics.

FACING THE QUESTION OF INDETERMINACY. In every type of theory we glanced at there remained some element on which the answer depended. It depended on the method of cognition of values, or it depended on the nature of God and the choice between prophets, or it depended on the structure of rational law and the meaning of terms describing actions, or it depended on the psychological structure of man or

15. Robert W. Leeper, "A Motivational Theory of Emotion to Replace 'Emotion as Disorganized Response,'" *Psychological Review* LV (1948), 5-21.
16. See above, p. 26.

the character of the human emotions, and so on. Now it is open to the absolutist defender of complete determinacy or the relativist defender of basic indeterminacy to argue that further examination of one or another of these dependency points settles the question once and for all. And this in fact is what both the extreme positions have done. Each has its characteristic armory of reliable phenomena, assumed fact, prescribed method to seal its case. Ethical absolutism has leaned heavily on the phenomenon of conscience, the sense of moral law, and the appeal to the light of reason. Ethical relativism has turned to individual will or attitude and focussed on the phenomenon of ultimate disagreement. And indeed if conscience does furnish decisive verdicts, if we do have a sense of moral laws as binding absolutes, if reason does lay down a definite line, then indeterminacy may prove ethically negligible. Or if the individual's will has an ultimate veto and if in ultimate disagreement there is nothing left to say, then ethical indeterminacy will be always with us.

These questions will accordingly have to be faced directly. The general hypothesis governing our investigation is that the dependency-points in the major ethical theories cannot be resolved without opening the doors to the fuller participation of the human sciences in facing the problems of ethics. The positions taken in the central assumptions of ethical absolutism and ethical relativism in their extreme forms have served as barriers against this fuller exploration, in part because they have tried short-cuts to avoid the necessity of empirical investigation, and in part because they have enshrined in their formulations the results of "commonsense" or of earlier chapters in the history of science. Yet these barriers require careful study, for underlying each there is a sound concern with some phase of the ethical process and the problems of method in ethics. In the next two chapters we shall therefore examine some central assumptions of the opposing positions. We shall find that the barrier effect tends to disappear when the methodological issue is clarified, and metaphysical pretensions and entrenched conventions frequently give way to vistas of empirical research.

II. Barriers and Tools: Conscience, Moral Law and Reason

§ Conscience as Pointer Reading

TRADITIONAL CONCEPTION. Among the ramparts of ethical absolutism none is more impressive than the appeal to the phenomena of conscience. "Conscience" is an old-fashioned word, long flayed by the relativists, but the phenomena involved readily cast themselves in other garbs. "Guilt-feeling" is scientifically respectable, "anguish" adds a poetic touch to anxiety, and of course remorse is familiar enough to outlive any sophisticated criticism. To consider appropriately the phenomena of obligation would be as great a task as the whole problem of ethical relativity. But shifting attitudes to its place in ethical theory are significant for the question of indeterminacy.

Traditionally conscience was given a _cognitive_ status; it was regarded as a source of moral knowledge, a faculty of acquiring it or an avenue through which it came. Moral philosophers debated about the kind of knowledge it gave us, whether universal rules or particular judgments. Some thought it could assure us that lying is wrong but not whether to regard a particular situation as one in which the central issue is lying. Others thought that universals were simply generalizations about the verdicts of con-

37

science; conscience could tell us here and now what was right, but we would then have to figure out why it was so. Socrates' conscience, as pictured in Plato's dialogues, is a rather more modest thing. It never told him what to do, and certainly never laid down rules; at most it made him feel uncomfortable if he was going somehow astray. In modern philosophy, Kant gave conscience a central philosophical role, comparing the grandeur of the sense of moral law with that of the starry heavens above, and suggesting that while the traditional theology had no real evidence that the voice of conscience is the voice of God, close attention to the phenomenon of conscience revealed our demand that the universe be a moral order.

CHANGING CONCEPTIONS. Changing conceptions stemmed largely from scientific inquiry into conscience as a phenomenon. Darwin took the moral sense to be part of the family of sympathetic reactions inborn in the human species, a fortunate acquisition in evolutionary history.[1] John Stuart Mill earlier, and Freudian theory later, took it to be a phenomenon acquired in the early years of individual life out of the pressures of the individual situation and familial-social relations.[2] In the 20th century conscience had a rather chequered career. Its eclipse in ethical theory came largely from the older anthropology and sociology which rested content with showing that different, often conflicting, moral judgments were to be found in different cultures. This was bolstered by the older behaviorist psychology which assumed such plasticity that in principle almost any set of moral rules could be inculcated. Contemporary psychology and anthropology have shown variety not only in the content, but in the very patterning of conscience. But instead of dismissing the phenomenon they have gone on to study the bases of its variant forms—guilt patterns, shame patterns, and so forth—in individual and cultural developmental proc-

---

1. In ch. 4 of his *The Descent of Man*.
2. J. S. Mill, *Utilitarianism*, ch. 3; Sigmund Freud, *New Introductory Lectures on Psychoanalysis*, trans. W. J. H. Sprott (New York, Norton, 1933), pp. 88 ff.

esses.[3] Such studies, together with the large role of guilt in contemporary psychological theory of personality development, have definitely ended the eclipse of conscience, although it may be difficult to recognize the somewhat austere, if not prim, voice of older days in her newly expanded domains.

RESULTING ESTIMATES. The resulting shift in the evaluation of conscience as a phenomenon is to see it as a mechanism with definite tasks in the development and organization of life. Julian Huxley, for example, thinks of it as, "a makeshift developmental mechanism, no more intended to be the permanent central support of our morality than is our embryonic notochord intended to be the permanent central support of our bodily frame."[4] Erich Fromm distinguishes authoritarian and humanistic forms of conscience, with the implication that societies can develop the latter type to some extent.[5] The ethical problem of conscience is therefore greater than ever before, since we realize that patterns as well as content have to be evaluated. And the direction of evaluation depends in large measure on understanding the psychological-cultural nature of the phenomena and what they render possible in human development. Some indications on these questions will be given in the chapter on *Psychological Perspectives* below.

CONSCIENCE AS A MEMBER OF THE FAMILY OF POINTER-READINGS. The methodological problem of how far the verdicts of conscience furnish determinacy in ethical judgment is, however, our primary concern. From this point of view, in the light of the shifting conceptions briefly outlined, it is clear that conscience constitutes no basis for an ethical

---

3. Margaret Mead, "Social Change and Cultural Surrogates," *The Journal of Educational Sociology*, XIV, 92-109 (October 1940); also "Some Anthropological Considerations Concerning Guilt," in *Feelings and Emotions*, ed. M. L. Reymert (New York, McGraw-Hill, 1950), pp. 362-373. For a recent critical analysis, see Gerhart Piers and Milton B. Singer, *Shame and Guilt* (Springfield, Illinois, Thomas, 1953).

4. T. H. Huxley and Julian Huxley, *Touchstone for Ethics* (New York, Harper, 1947), p. 252. See also p. 150 below.

5. *Man For Himself* (New York, Rinehart, 1947), pp. 143-172.

absolutism. Even attempts to give guilt feeling a cognitive theological status will have to rest independently on their theological theory rather than on the character of the conscience phenomenon.

Methodologically we may best conceive of conscience as providing pointer-readings. Guilt-feeling, shame, as well as any other feeling-reactions that may be brought into the range of moral "tests" either by anthropological comparison or ethical theory (for example, pride and fear, or sympathy and indignation) play a role comparable to familiar operational tests in science. The pointer on the scale "tells" us the weight, the change in color of litmus-paper "tells" us whether the liquid is acid or base, the noise of the alarm-clock "tells" us it is 7 A.M., the thermostat setting "tells" us the temperature conditions under which the furnace will go on. Obviously we are not literally told in each case. The meaning of the pointer-reading comes from our assumed knowledge of the field that is involved together with our knowledge of the mode of operation of the measuring-device. Similarly, what guilt-feeling "tells" us about the moral field depends in large part on our assumptions about man's nature and development and the origin and functioning of guilt-feeling itself in these processes.

If the comparison to the physical science model seems remote, intermediate models can be found along the range of the sciences. Guilt-feeling may be compared to a medical symptom. It shows that something is wrong, but precisely what depends upon the knowledge of the disease, and of the patient's medical history and physiological condition.

That the family of so-called "moral feelings" is admirably suited for the role of pointer-readings in a systematic ethical theory seems indicated both by the history of ethics and by the results of the study of personality and emotional development today. Both differ markedly from the disregard of feeling-reactions and emotional reactions as secondary and subjective which characterized a transition period in the psychological sciences. In the history of ethical theory there is continual grappling with fine distinctions: for example,

Butler's attempt to separate self-love from particular passions, and both from conscience; or Kant's insistence on differentiating feelings of conscience from fear of consequences or self-praise and self-blame, or again, his distinction between self-love and self-esteem; or Adam Smith's elaborate pyramiding of sympathetic reactions to sympathetic reactions.[6] And in contemporary psychoanalytic psychology, the subtlest shade of feeling is readily seen as a sign or indication pointing beyond. But the feelings do not tell their own story, and their meaning can grow only with the growth of systematic theory of man and his ways.

§ "The Starry Heavens Above and the Moral Law Within"

TRADITIONAL CONCEPTIONS. The sense of moral law is the second bulwark of ethical absolutism that we are to consider. Kant's coupling of this marvel within with the marvels overhead involved a complexity seldom equalled within the confines of a single moral concept. There is the binding peremptory quality of conscience, the universalizing tendency of reason with its authoritative stamp, the emotive association of human legality and physical regularity, as well as the energy latent in specific historical forms that the concept had taken.

The universalizing tendency reached special heights in the 18th century. This has been attributed to sources as varied as the growing effect of Renaissance humanism in man's discovery of his essential humanity, the breakdown of localism in the discovery and exploitation of a new world, the emancipation of the individual in the religious transformations, the tendency of the bourgeoisie as a rising class to speak in the name of all mankind, the influence of a

6. For Joseph Butler, see especially his first sermon (Little Library of Liberal Arts edition, pp. 19-32). A good example of Kant's analysis is in his *Lectures on Ethics* (translated by L. Infield, London, Methuen, 1930), pp. 129-35. For Smith, see his *Theory of the Moral Sentiments* (London, G. Bell & Sons, 1911) part III.

growing world trade, the fascination of the physical model in Newtonian universal form. Even apart from rationalistic emphases on moral laws, metaphysical assumptions about nature or divinity had played and still play their part in magnifying the conception. Moral law could be represented as expressive of man's nature, as Aristotle took natural justice to be. Or it could be seen as a reflection of divine reason, as Aquinas believed it to be. To these have been added the view of moral laws as axioms of the moral life, a standpoint which many an ethical theorist with a lingering respect for Euclidean certainty still stoutly affirms. Natural order, natural law, natural rights, divine order and divine law, rational order and rational axioms all combine to enhance the conception of the moral law. Accordingly moral laws tend to appear wearing a crown of absolutism and brandishing a sceptre of necessity. And their garments are resplendent with a special must-ness.

TYPES OF MORAL LAWS. If we put aside all background issues and concentrate simply on the concept of moral laws and the qualities associated with them, we find that rather than a single form we have a whole family of forms. Different types of moral rules are found in ordinary usage, such as "killing is wrong," "most killing is wrong," "killing as such is wrong," and so forth. (Similarly for rules employing "duty," "right," etc.) But rather than collect examples of such variety we will perhaps get to the heart of the question more quickly if we ask what feelings about moral rules emerge in contemplating decision-situations. Technically, this would be called an inquiry into the *phenomenology* of moral rules and it would not strictly concern how we *feel* about them, that is whether we feel happy or worried or secure, but what qualities they have in the field of our "vision."

At least five sorts of cases can be distinguished. There are *must-rules* carrying a mandatory quality, brimful of necessity, with a bitter "woe unto you" taste. There are *always-rules* whose universality betokens a reliable dependency rather than a menace; you feel about them as you feel

about pressing a button to switch on the lights—they tell
you exactly the right thing to do. Then there are the *break-* ✗
always-with-regret rules; they have an exalted status—you
can see the gleam even from afar—but they do not have an
absolutely compelling power over the individual situation.
Beyond these lie the *for-the-most-part rules;* they carry a ✗
certain authority, but where you find them inapplicable you
feel no qualms in departing from them, since you are not
really "breaking" them. And finally, there is the limiting
case, the statement depicting a *complex singular situation,* ✗
in which no rule may introduce a binding element. Each
type merits separate consideration.

MUST-RULES. The must-rule is familiar enough both in ①
human history and in ethical theory. It characterizes some
types of action as morally imperative in the most emphatic
and unqualified sense. It refers to any and every situation
in which its subject applies, but it seems to involve more
than simply universality. The familiar religious view of
some sins as utterly and irretrievably damning illustrates
this type. The Victorian attitude to unchastity is an example.
Pacifism embodies a similar attitude towards killing. Gandhi's non-violence principle made him have scruples even
about killing dangerous snakes. And it included non-resist-
ance even under extreme conditions.

The standard philosophical illustration of this type of
approach is Kant's use of his categorical imperative. The
heart of his outlook is clearly expressed in the reasons he
gives for his argument that you must not lie even to a
would-be murderer about his victim's whereabouts.[7] For he
goes on to say that if you tell a lie and if by chance the
victim meanwhile went in that direction and was killed, the
fault would be yours since it followed your violation of the
moral law. But if you tell the truth and the victim is killed,
the fault is apparently nature's. Your hands are clean. Purity
of individual soul thus emerges as Kant's implicit goal, as

---

7. "On a Supposed Right to Tell Lies from Benevolent Motives," in
Abbott, *Kant's Theory of Ethics,* pp. 431-437.

it does Gandhi's. And Kant seems to have the added assumption that nature is so thoroughly unreliable one had better not try to act on probabilities.

The question whether the must-like quality attaches to a rule as such or is a separable phenomenal quality, which may be detached and pinned on elsewhere, is an issue of paramount importance. Kant does not even raise the question, because he takes the very phenomenon of conscience to be that of an internal tribunal judging according to law. But if this constitutes only one type of conscience-pattern, if other styles of conscience are possible, then whether the must-quality is better off in one place or another becomes a question of evaluation. Bergson, I think, sensed the problem in his comments on the assimilation of the ideas of law and command: "The two ideas coming against each other in our minds, effect an exchange. The law borrows from the command its prerogative of compulsion; the command receives from the law its inevitability."[8]

From this point of view it is important to note that the quality is not limited to a universal which serves as a premise in a moral analysis; it may characterize a universal that happens to be a conclusion *after* the analysis is performed. And the conclusion may recommend a line of conduct far different from that of the usual moral maxims. Take, for example, a statement issued by the French resistance during the Nazi occupation, headed "The Duty to Kill."[9] The obligation statement is universal in form, addressed to all Frenchmen. It examines alternatives—"You have no choice: either you will come back into the war or you will perish." It describes the temptations to hold back, the evils inflicted on people, the rationalizing that becomes surrender. "What is the answer to these demands and to these methods? Only one decision is possible: KILL." Even a morality that would suffer in silence should not accept evil done to the whole

8. Henri Bergson, *The Two Sources of Morality and Religion*, translated by Andra, Bereton and Carter (New York, Holt, 1935), pp. 4-5.
9. In A. J. Liebling, editor, *The Republic of Silence* (New York, Harcourt Brace, 1947), pp. 430-432.

country. It has become a whole-life battle. "WHICH DO YOU CHOOSE, LIFE OR DEATH?" The editor adds a quotation from Vercors: "By his abominable acts, the enemy has made hatred almost a duty."

Now letting go the question of the correctness of this moral judgment, it may be argued that in this example we do not have a genuine universal, but only an application to a particular situation in which all Frenchmen happened to find themselves. Certainly it is not a universal law in Kant's sense. A number of alternatives are therefore possible in analyzing its imperative quality. The conclusion may be a "theorem" deriving its must-like quality from more general must-laws, together with factual statements about existing conditions. Or it may be quite simply that the mandatory quality characterizes *any* particular decision in an important area where a mass of evidence has shown that all other paths are cut off and only one is left clearly determined as the line of duty. If this is so, and if even a singular statement of duty may be associated with the phenomenal quality, the connection between "must" and "moral law" is not an inevitable one. We would therefore have to look for its causes. This appears to be a psychological problem. If we acquire our obligation feelings from internalization of parental commands, then it is understandable that these should take a generalized form and moral compulsion be felt as tied to moral law. But adult recognition may separate them once more. And then it becomes relevant to evaluate the locus of the must-like quality, as well as its inner nature and content. Fundamental policy decisions have to be made. How strong a sense of compulsion is desirable? Should it attach to universal rules or is that too rigid a morality in human life? Should it attach instead to the particular decision, or does that atomize human life too much? Have we enough knowledge of man as yet to answer such questions? At any rate, the questions are there, and policy decisions one way or another have to be justified.

ALWAYS-RULES. Now then, what about *always-rules?* They would be universal moral rules of an empirical character. They take the form "In all cases where a definite set of circumstances *a b c* occur, it is my duty to do *d*." The type is clear enough from rules in handling technical processes: e.g., "Whenever the temperature indication is above so-and-so many degrees, turn off the machine." Why turn it off? Because the machine will wear out, or something is wrong, or there may be an explosion. Such rules have a large component of instrumental value, but if goals and conditions are stable enough they may nevertheless remain as dependable universals. Their prevailing quality will be more that of *appropriateness* to a standardized situation than of must-ness. Perhaps the clearest examples are to be found in relatively separable duties intrinsic to certain enterprises, as in the ethics of professions or occupations. Firemen must be on the alert, workmen in public transportation exercise constant caution, doctors in practice keep up with newly-developed cures, and so on. All such general rules derive their moral quality from the importance of the enterprise and the values they support, or the seriousness of the dangers they avoid. Occasionally even major moral rules have been cast in such a form—for example, when honesty is recommended as the "best policy" and the only safe course on every relevant occasion.

BREAK-ONLY-WITH-REGRET RULES. These seem to me to constitute a central type insufficiently analyzed, whose careful exploration may resolve some moral dilemmas. They are probably the type intended in such formulations as "Killing *as such* is wrong, but it may be your duty to kill on a specific occasion." The essence of such a rule is that one respects or holds it in the very performance of the action condemned by it. If the rule were a must-rule, such performance would be a moral violation. If the rule were an empirical universal, such performance as a moral choice would be refuting it in a test case. Hence, if the break-only-with-regret rule remains unrecognized, one slips into the formulation "Most killing is wrong" instead of the general "Killing is wrong." The establishment of the break-only-with-regret rule as a distinct and separate possibility is therefore of considerable theoretical importance, especially as it seems to correspond with the picture as most people see it; for I suspect most people wish neither to argue that every actual case of killing is unjustified, nor to deny there is some sense in which killing is wrong as a universal.

A possible way out is to regard a break-only-with-regret rule as referring not directly to the *results* of moral calculation of a situation but to the *process* of reaching such results. Hence the rule is what we may call a *phase rule* or states an _operational universal_ or rule of reckoning. What it says is that in every situation in which a moral reckoning is taking place, the killing aspect is to be reckoned as negative (wrong). This is a universal statement. It does not guarantee that the outcome of the reckoning (one's duty in that situation) will be negative in the sense of excluding the act.[10] Thus "Killing as such is wrong" tells us that where-ever the situation includes a killing its value will be much less than it would have been if, other things being equal, the same relevant result had been secured otherwise; but it does not say that the act of killing cannot under special circumstances rescue greater values (e.g., many lives). It may even then be one's *duty* (based on the results of the reckoning) actually to kill or risk killing, as in certain war-time situations.

The phase rule therefore is a genuine universal, analyzing one abstract aspect of moral situations. Whether it is empirically established is a separate question; but in principle it may be seen as comparable to the scientific process of tracing the consequences of one force under ideal conditions in abstraction from the particular complex situation, e.g., the part of an object's motion that is due to initial impulse as distinguished from the diminution due to friction.

Such an analysis is clearly in accord with many facts in moral experience. For example, we differentiate between telling a lie while respecting truth and simply telling a

---

10. As an analogy, subtracting a thousand from any positive finite number always makes the total less than it would have been, but it does not always make the result negative.

lie. In some of the countries occupied by the Nazis in World War II, it was necessary to retrain children who had grown up under the occupation skilled in lying to the Nazis, not to use this instrument after the liberation; presumably this would not be necessary for an adult who felt pained and degraded by having had to lie.[11] Similarly, respect for life is especially necessary for those who are in a position in which they may have to kill, if they are not to become brutalized. Respect in such cases need not constitute hypocrisy; although it will be such if it be only lip-service to the ideal.

FOR-THE-MOST-PART RULES.   These are simply probability statements indicating (roughly) the frequency with which a type of action will turn out to be one's duty where relevant, or a phase-element in reckoning. In neither case is it binding rather than informational; any binding element comes simply from any obligation to act on probabilities where there is insufficient analysis of the particular. The central point phenomenally is that there is no regret where the rule is abandoned in a particular case because duty is found to lie in another direction.

Examples of for-the-most-part rules in ordinary morality concern types of actions that are important in some but not all fields, and are usually understood as such. For the most part one ought to *persist* in pursuing a goal; but where one does not it need not be with regret. On the whole it is wrong to exhibit anger; but it may sometimes be perfectly right to do so. Punctuality requirements vary in different types of social relations. While some of these may tend to become phase rules, they are not always, and there is no a priori reason why they should have an inevitable plus or minus in all moral reckoning. For-the-most-part rules have

---

11. Vercors, in the passage cited above, says addressing the enemy: "Of all the reasons for hating you, O you that I cannot call my fellow men, this single one would suffice: I hate you because of what you have made of me. Because you sowed and cultivated in me, with the diabolical persistence that is yours, sentiments for which I can feel only disgust and scorn." (Liebling, *op. cit.*, p. 432).

a definite reference to place, time, and social and cultural conditions, as well as prevalent circumstances. But there may be some unqualified rules of such a sort for all men at all times.

UNIQUENESS DUE TO COMPLEXITY. Finally, it cannot be ⑤ assumed a priori that moral rules of the above types are to be found for all moral situations. Singular moral statements may alone be possible in complicated fields in which decision seems to require a kind of synthetic grasp which cannot be expressed as a reckoning of distinct aspects. At least the possibility of such must be admitted. For example, problems of personal affectionate relations seem so intricate that the most one may demand may be an emphasis on some general virtues such as sincerity and insight. These are the kinds of cases in which one is often told simply to follow one's conscience, on the assumption that conscience is a N.B. synthetic faculty which puts together the moral claims of the particular situation in a single comprehending judgment. Philosophers who lay most stress on complex particular situations are prone to look down on rules as static abstractions, and to underscore the uniquely creative character of man in moral decision.

CONCEPTION OF MORAL LAW DOES NOT ENTAIL MORAL ABSOLUTISM. If this whole analysis of moral law is on the right track—and it does seem to me to be dictated by contemporary logical analyses of "law" as well as by psychological and anthropological treatment of the "must" element—a moral absolutism can no longer rest assured on the sense of moral law as on a rock. We need not explore at this point the frequency of occurrence of the different types in western morality—for that would not be decisive, it would simply describe our particular historical moral pattern. Nor need we now evaluate the different types by asking whether one or another should in general be dispensed with or enshrined. Nor can we assume that the analysis yields an ethical relativity automatically; for it might turn out to be desirable to employ each of the five types in different areas of human life under a value scheme that can yield definite and correct answers. Even the desirability of must-rules of the most Victorian vintage is not precluded a priori. A reasonable decision of its utility requires a knowledge of the role of guilt-feeling in the development of desirable types of personality. It requires a knowledge of the alternatives possible in the feeling-mechanisms of development. It requires consideration of what absolutes modern society may require and what psychological mechanisms can secure them. And no doubt it requires answers to many questions not as yet clearly formulated. On the whole, a liberal civilization from the time of John Stuart Mill at least, has hoped that it will be possible in the long run to replace peremptory guilt patterns with aspiration patterns embodying a vital sense of obligation. And there is significance in the degree to which peremptory guilt patterns are associated with authoritarian institutional forms. But in our present context of inquiry it is enough to see that on analysis this central fact of ethical absolutism unrolls into assumptions of fact requiring scientific investigation and decisions of policy requiring further evaluation.

§ The Romance of Reason

THE APPEAL TO REASON. The conception of reason has been closely intertwined with those of conscience and of moral law in attempts to enshrine a moral absolutism. Nor has it altogether lost its hold; to appeal to reason is still a primary mode of justifying a course of action, and likely to remain so. Our purpose is not to add to the scourge of irrationalism by unseating reason, but to find its fuller meaning and the source of its authority. Our guiding hypothesis is that it comes from the accumulation of results in human inquiry, not from some uncanny faculty of penetrating the mist by which mankind is enveloped.

The basic phenomenon underlying all consideration of reason is the fact that men think. Underlying the claims of reason at any time we may therefore expect to find the current analysis of the nature and role of the thinking process in the human being. In some sense then, the validation of reason is a function of the existent theory of psychology, although in a complicated way, since thinking itself plays an integral part in building up psychological knowledge.

The correctness of this general approach to reason, the way in which the concept remains a barrier as long as it is absolutized, and how it becomes a valuable tool in the effort to achieve a greater determinacy in ethical judgment when its role and contents are made clear, can best be seen in brief compass by a survey of the romantic history of reason in human thought.

REASON ENTHRONED. A preeminent place is given to pure thought in ancient philosophy. Plato's Ideas are grasped by pure intellect which rises above mere sensory perception; absolute moral truth comes from grasping the Ideas of justice, courage and so forth, culminating in the Idea of the Good. Aristotle's god is pure thought thinking about itself. In major mediaeval trends, reason is almost (although not quite) coordinate with revelation; at least she is adequate to grasp the natural law and see the basic structure of human duty and obligation.

With the growth of mechanistic science reason added to her stature. She provided the theoretical axioms of moral knowledge in a world that was taken to be ultimately as systematic as Euclidean geometry. The Age of Reason proclaimed her natural light sufficient to sweep away the accumulated customs that enslaved mankind, and at the most critical point of the French Revolution Robespierre erected a shrine to the deity of Reason. Even when the revolution was over and reaction appealed to historical categories, she reappeared in Hegel as the grandiose Idea providing the structural framework for the whole of the world and history. By her versatility, by establishing profitable connections with varying interests, she maintained her throne—until the age of intellectual monarchy went by.

DECLINE AND FALL: REASON AS SERVANT OF THE PASSIONS. Yet even while the Age of Reason was growing in France, in England Hume prefigured the fate that was to befall her,

and assigned her the role of servant of the passions. This came to pass with the strengthening of empiricism in the sciences, and the emergence of influential positivist philosophies. Reason could no longer legislate in morals. She could but wait for the selection of ends and then, keeping her eye on the goals proposed, faithfully reckon the means to achieve them.

There was no end to the indignities heaped upon her. Her motives were questioned and her judgments ridiculed. How is it possible after Freud and Marx, it is still said, to have any confidence whatsoever in reason? At any time it may just be working to produce rationalizations for deep controlling forces. And rational ideals, however luminous, may be covering the grossest of class interests, with their very purity simply a garb for exploitation. Hence current works in inspirational pessimism have reached the point of revelling in the twilight of civilization, and parade Dostoyevsky as the apostle of anguish, supplemented by the existentialists; they serve a dish salted with the sociology of knowledge and peppered with Pareto. All in all, a lively celebration for a funereal setting.

VIRTUE TRIUMPHANT: THE REEMERGENCE OF REASON. Our heroine, with a well-grounded sense of her worth, refused to perish. She bowed her head before the storm and busily plied her menial tasks. She had already taken first steps by entering into a stable marriage with Experience, whom in her heyday she had herself regarded as a slave. Henceforth it was the teamwork of reason and experience that counted. Her husband foraged for facts, she sorted and systematized, spinning her web of theory. She made nets for him to fish with, and maps for him to find his way to larger and larger quarry. Accepting the lesson that her creations were not direct insights into reality but outlines of alternative possibilities in carefully fashioned systems[12] she found a substitute satisfaction for dogmatism in the freed creativ-

12. Most strikingly established in geometry by the development of non-Euclidean systems in the 19th century and their use in physics in the 20th.

ity of her fancy. Her husband too outgrew his slavish adherence to isolated facts and the belief that he could build a home of knowledge by piling them one on top of the other. Eventually their teamwork became so successful that it was hard to tell where the work of one ended and that of the other began. More and more when men talked of reason they meant the team and they even began to use increasingly her married name of Scientific Method. For a time the family became puffed with pride and began to impart to Scientific Method some of the omnipotence that had characterized the divine right theory of Reason in her youth. But they are slowly learning better.

Working together reason and experience do not need to fear that they will be undermined by irrationalism. For they know that irrationalism overlooks the implications of its own sources, that the weapons of abuse come from the labor of the very objects of abuse. If the Freudians have shown that reason is rooted in the development of instinctual forces, they have also studied the development of these forces in the fashioning of personality. And so they provide us with tools for distinguishing pathological from non-pathological or realistic forms of expression.[13] And if the Marxians have provided tools for digging out the narrow social content of ideals and philosophical conceptions, then such tools are also available for seeing what elements of permanence, of all-human reference there are in the procession of ideals on the face of history. So far from overthrowing reason—except in its pretension to provide an absolutism on its own—modern developments in the psychological and social sciences have explored its roots and furnished criteria for separating the wheat from the chaff.[14]

---

13. See, for example, Heinz Hartmann "On Rational and Irrational Action" in *Psychoanalysis and the Social Sciences*, edited by Geza Roheim (New York, International Universities Press, 1947), vol. 1, pp. 359-392. With respect to a proposal that the concept of the "irrational" be dropped, he says: "We refer to rational and irrational as to empirical psychological characteristics of action that may be present or absent. In this sense the terms are meaningful and useful" (p. 365).

14. For a fuller discussion of this point in relation to the objectivity of truth, see below, pp. 282-287.

THE PRESENT SITUATION. On the whole the Scientific Method family has been thriving busily. And it is only now reaching the point where it can try to reestablish some of its claims to a greater part in rendering ethical judgment less indeterminate. It has been too overawed by its confinement to the servants' quarters to see how much its scope has really been enlarged. It has accepted the disavowals of social scientists that social science is not concerned with ends or values but only with means and the factual question of the maximization of goal achievement.[15] But simply to choose the best means for an assigned end—and more so when several ends are concerned—already involves value considerations. If one is told to figure out the costs of achieving the end, attention must be fastened upon the consequences. A "neutral" mapping of consequences includes their effects on the person who assigned the end, hence an investigation of the degree and quality of his attachment to his proposed goal. For it is elementary that a person's whims are to be distinguished from his stable and enduring goals, and that a rational man who is to be regarded as master of himself must have some insight into himself and his aims. But once this door is opened, the passageway leads on and on. Ideally, full advice to the person who asked for a cost estimate of the envisaged goals would include their scope and function in his life, their mode of development, intensity points, termination points, possible transformations, relation to common ends, possible interactions with ends of others and their mutual alteration, and so forth—in short, all the lessons that a psychological, social and historical perspective could offer in application to the particular case.

Such an expanding conception of the role of reason need no longer limit itself to the model of the messenger boy carrying a sealed parcel to be delivered to a person at whom

15. For fuller discussion of the fact and value dichotomy, see below pp. 74-77, and pp. 106-109; of value judgments in social science, pp. 242-246; of maximization of achievement, pp. 62-64.

he does not give a second glance. It can use the model of
the expert advisory commission instructed to ascertain "all
the facts," preparatory to legislation, on the state and pros-
pects of natural resources and their utilization in the light
of existent conceptions of public welfare, which they are
also permitted to explore! But this mandate does not restore
the tyranny implicit in the older sovereignty of reason. The
commission must exercise no coercion, and it does not vote
in the legislature.

Such an altered conception of rationality in relation to
values does not determine in advance how far it will be
able to go. It simply removes a priori barriers to its expansion. The earlier dominance of reason reflected a psychological theory of a pure energy grasping an ideal structure, and it was undermined by conceptions of thinking as
a manipulation of sensory elements. The reemergence of
reason reflects a theory of the active functional role of
thinking in relation to biological and psychological processes in a cultural-historical context. A full contemporary
account of reason would need to explore all its components
and their relations. It would not rest content with an exposition of logical theory and its ideals of consistency and
system. It would also analyze the way in which logic combines with experience to yield progressively more stable
knowledge. And it would show what purposes of men enter
into this process, what unity it imposes on their lives, how
it stabilizes its authority, what leading conceptions it forces
on men's thought, what attitudes it engenders and requires
for its persistence and growth, and what opportunities it
offers in human life. Let us see how ethical theory utilizes
these concepts and how far they carry us in narrowing down
indeterminacy in ethical judgment.

§ Contents of Reason in Ethics

CONSISTENCY AS A MINIMAL REQUIREMENT. The first and
minimum requirement is that we be consistent and not
contradict ourselves within the process of ethical judgment.

Now how far will this requirement actually carry us in ethics? Is it one that almost any moral system can easily satisfy? Or does it shape the form and content of morality into some determinate pattern?

Kant worked out a whole ethical scheme in the name of consistency. The procedure he propounds is to decide the morality of a course of conduct by taking the particular maxim underlying the action and seeing whether it can be consistently willed, which he takes to mean universalized. Suppose, for example, Kant says in his *Fundamental Principles of the Metaphysic of Morals,* that a man finds himself forced by necessity to borrow money which he will be unable to repay. But he knows that without a promise to repay in a definite time no one will lend him the money. If he decides to make the promise, then the maxim of his action is that when one thinks he wants money, he will borrow and promise to repay although knowing he cannot do so. This may be consistent with one's own future welfare, but what if the maxim were to become a universal law? Kant, at least, sees at once that it would necessarily contradict itself: promises would become impossible, no one would take them seriously, and woe unto the needy but solvent would-be debtor as well. In other kinds of cases, such as preferring indulgence in pleasure to developing one's capacities, Kant sees that such a system of nature could indeed exist but a man cannot possibly *will* that this should be a universal law of nature; Kant believes a rational being cannot will the non-development of his faculties.

Kant's reliance in these examples on the ruinous consequences of deceitful promises and on what a man cannot possibly will, shows that he is dealing with more than purely logical consistency. His universalization test itself cannot determine moral content without some evidence of what men might actually be ready to universalize. In another context Kant cites the case of perpetual war among Indian tribes which has no other object than slaughter, even adding that "bravery in war is the highest virtue of savages, in their opinion."[16] But instead of calling it moral because they have willed it universally as a virtue, he offers it as an example of depravity. One is tempted to paraphrase Russell's remark about other philosophers in a parallel context, and say that Kant sets up an erroneous criterion of evil and then finds any man really depraved whose act of will is such as to prove Kant's criterion wrong.

But all this only points to the difficulty in using mere consistency in judgment as a basis for fixed value content. It may, in fact, be doubted whether Kant really intended just this. The very notion of consistency as a necessary component of rationality must have something more than logic in mind if it is to reduce indeterminacy in ethical judgment. That something more is clearly either a value element or a special lesson of experience. If so, it should not be smuggled in, but be registered with name, address and credentials..

ATTEMPTS TO DEVELOP A CONCEPT OF PRAGMATIC CONSISTENCY. Some philosophers, recognizing that consistency in ethical judgment means more than purely logical consistency, have tried to develop explicitly a notion of *pragmatic consistency*. The additional element is the demand that in ethical judgment one be far-sighted rather than simply think of the moment, that one's whole life weigh as much in deliberation as the present or nearer future, that one's attitude be serious rather than drifting, that life be oriented to avoid having anything to regret. Such a conception was presented by C. I. Lewis in his Carus Lectures some years ago.[17] He does not, however, cast the additional element as a value or a lesson of experience, conceiving it rather as a fundamental categorical imperative ("Be consistent in valuation and in thought and action," "Be con-

---

16. "Of the Indwelling of the Bad Principle along with the Good, or On the Radical Evil in Human Nature," in T. K. Abbott, *Kant's Theory of Ethics* (London, Longmans, Green, 1879), pp. 401-402; cf. p. 397.

17. *An Analysis of Knowledge and Valuation* (La Salle, Illinois, Open Court, 1946), ch. XVI.

cerned about yourself in future and on the whole") of which he says: "It requires no reason; being itself the expression of that which is the root of all reason; that in the absence of which there could be no reason of any sort or for anything."[18]

But why should a man decide that his whole life is to be regarded as one unified life rather than a succession of relatively independent seasons, or as a stream with only a general continuity, or even (to reverse the direction) as a partial component in the collective life of his family, his society, all mankind? There may be a sound basis for a whole-individualism as against fragmentized-individualism or some type of collectivism. But the basis must be some theory of the world and its possibilities, some account of the psychological, social and historical nature of man.

Again, why should regret be avoided at all costs? Why not do some things we will be sorry for later? Why make the fact of regret the firm core of rationality in ethics? And can it be avoided only by a serious plan of life? Is that because it is a psychological fact that one cannot simply erase one's feelings about the past, and they cause pain and internal conflict? This value core in the notion of pragmatic consistency leads us back once more to the psychological nature of man.

It is very probable that the minimal meaning of rationality in ethics—the notion of consistency in action as against purely logical consistency—involves as its value core the attempt to minimize internal conflict and avoid the sense of frustration. It is clear that there is no logical inconsistency in holding two opposite values together; for the phenomenon of ambivalence has long been recognized. Plato, in fact, uses the phenomenon of the man wanting to do something and yet not to do it at the same time as proof that there are different parts of the soul underlying the separate tendencies. If ambivalence embodies an ethical inconsistency it is because internal conflict produces pain or frus-

---

18. *Ibid.*, p. 481.

trates whole-hearted choice and inhibits action. The root value judgments in a minimal conception of rationality that aims to lessen indeterminacy are thus that pain is an evil, that frustration is evil, that action is required in human affairs. These judgments and their warrant in human life will be analyzed more carefully below.[19]

UNITY. The insistence that rationality involves some systematic unity of the person as against the disparate functioning of the "parts" of the self is a common feature of appeals to reason in ethical judgment. It was not seriously questioned until Hume's treatment of the mind as simply a theatre in which impressions come and go. But Hume was not denying whatever *empirical* unity might be discovered in the operations of man, nor that different qualities might be sought in that unity. From to-day's psychological perspective it is established theory that there is a *system of forces* in the internal economy of the biological individual, even where consciousness shows the extreme of a split personality. In that sense it is a lesson of science that rationality involves the whole person, This serves to incorporate at least the whole-person perspective in ethical judgment. This history of the moral mandate of unity and system is itself one of the most striking cases of the tremendous scope of scientific knowledge that is really required to justify what on the face of it looks like an obvious injunction of reason.

AUTHORITY. An authoritative quality was early assigned as an element in the concept of rationality. If accepted, it would mean that in ethical judgment whatever stemmed from the intellect would be objectively superior to what stemmed from other components in the human makeup. Thus Plato took reason to be the authoritarian dominant element in the soul, ruling over the passions and the will.[20] He regarded passions and appetites as blind and capricious, striving only for satisfaction, therefore requiring the constant repressive veto of reason. A priori, he may be correct. But

---

19. pp. 132-137, 144-148.
20. See, especially, Book IV of his *Republic*.

equally, he may be wrong. The picture he gives of appetite may be simply the picture of its revolt in a system of repression, and the crucial factual question may
be to what degree this repression is unavoidable. Plato's
account becomes suspect because he uses it to justify a
regimented society in which reason resides in the rulers.
But his psychological picture requires independent factual
examination. The question whether rationality is to be construed as having an autocratic character or a democratic
character is in part at least then dependent on the results of psychological investigation. It is still being debated,
although in the language of Id and Superego in relation to
Ego rather than in Platonic terms. Whatever determinacy
an element of authority in the appeal to reason may add
in ethical judgment thus depends on the outcome of this
psychological investigation and the value assumptions it
probably entails.

There is, however, a general sense in which the appeal
to reason does carry some authority. On the conception of
rationality constructed above it means appealing to the
results of experience as best systematized by theoretical
reflection. Such authority stems simply from the weight of
experience. It is not to be identified with the functioning of
one part of the person as against another, as in the traditional authoritarian conception.

HARMONY.   Even the narrowest conception of rationality as the choice of means most conducive to ends held or
assigned is faced with the problem of reconciling the several ends. Hence, the idea of reconciliation, compromise,
harmonizing, is frequently found as a component of being
reasonable. Such terms often carry a strong emotional
charge,[21] and their inclusion in the rationality concept often
gives it a marked value tone. Attention must accordingly

---

21. See, for example, Margaret Mead's point that compromise is a
good word in British attitudes, but a bad word in American, in "A case
history in cross-national communication" (in *The Communication of Ideas*,
edited by L. Bryson, New York Institute for Religious and Social Studies,
1948, pp. 223 ff.)

be paid to the underlying model in terms of which harmony is conceived.

The classical conception of harmony as a component of ethical reasonableness was cast more in terms of artistic perfection than in those of striking an agreeable bargain or efficiently maximizing gains. For Aristotle,[22] the "man of practical wisdom" whose sensitive judgment of what is to be done embodies the "right rule," acts almost like an artist or craftsman sensing the proper combination of the raw materials according to time, place, relations. Thus the virtue of courage blends fear and confidence so that it becomes neither timidity nor rashness, and a social virtue such as liberality harmonizes giving and taking to yield neither prodigality nor miserliness. In its minimal sense this involves an initial acceptance of human drives as contrasted with a permanent value tabu on any one of them; they become regarded as raw material for the construction of life and character.[23] A somewhat stronger form is found in modern uses of the harmony notion as the core in a conception of reason, which extend this acceptance of the raw materials of the human make-up into an initial liberalism to any existent impulse, desire or appetitive tendency. Santayana's *Life of Reason* is a good example. That this need not appear in the conception of reason alone is clear on comparative grounds; the same value judgment is found in R. B. Perry's conception of *love* as "an interested support of another's preexisting and independently existing interest,"[24] or in Royce's notion of loyalty to loyalty wherever it is expressed and whatever its object.[25]

Going beyond these value judgments, the harmony conception often becomes a bias towards smoothing out conflicts between existent values and interests rather than

22. *Nicomachean Ethics*, especially Books II-IV.

23. For the evaluation of expression of biological drives, see below pp. 137-143.

24. *General Theory of Value* (New York, Longmans, Green, 1926), p. 677.

25: Josiah Royce, *The Philosophy of Loyalty* (New York, Macmillan, 1911), III.

rejecting one side or another in a sharp stand. Thus in Aristotle's *Politics* the notion of the mean emerges more as a *balance* between existing social forces of rich and poor, expressing themselves politically in the struggle of oligarchy and democracy. The mean seeks political forms balancing their forces so as to avoid conflict and revolution. In such conceptions of rational harmony, however, the value center of gravity turns out often to lie not in the demand for harmony but in the kind of chord to be struck. There is a tremendous difference—in harmonizing science and religion, for example—between saying that science may be free so long as it does not interfere with religious values, and saying that religion may retain such of its beliefs as do not conflict with scientific progress.

It follows that while rationality in ethics does include a strong tendency to minimize conflict, it has an ultimate hypothetical element: it does not of itself determine in which direction one is to go where an incompatability of basic direction is revealed.

MAXIMIZATION. In conceptions of rationality which stem more from the business model, the idea of a reasonable choice of means to reconcile a multiplicity of desired ends is given a more positive expansive turn. The idea of selecting means wisely to produce maximum net gain comes to the fore, embracing both the meeting of maximum demand and the minimization of cost or loss. Such an ideal of rationality, brimming with the spirit of accountancy, was found in the Utilitarianism of Bentham, with its outlines of a felicific calculus. It was tied to the assurance that happiness was the good and more happiness the better. In the development of economic theory since that day the ideal retreated from whole-life scope to the more limited domain of economic welfare. Here there have flourished a number of concepts constitutive of the idea of rationality as maximization.

Of these, with respect to means, *efficiency* is the most familiar. Means are used most efficiently when ends are maximized; and conversely an end is achieved most efficiently when means are minimized. This holds when the relevant ends and relevant means are all clearly distinguished, together with their consequences. In practice, the chief ethical danger is the neglect of collateral ends and consequences simply because they do not come clearly enough into the range of economic costs. But it is usually made clear that economic efficiency need not mean ethical desirability. Such a constituent of rationality adds determinacy to ethical judgment only within the means-ends relation. It receives the assignment of ends from outside.   *W. B.*

The concept of a *social optimum* as inherent in rational action has been developed in the field of welfare economics. This indicates, roughly speaking, a condition in which no one can be further benefitted without someone being harmed.[26] But in spite of the mathematical form and precision in this concept of economic welfare, the indeterminacy points remain transferred to special problems of technique and provision of value judgments required for application. For example, ethical assumptions about the desirable distribution of income are required if the theory is to avoid what Boulding calls "the accusation that some welfare economists assume the permanent vesting of all vested interests."[27] But fundamentally, the theory does not enter into the evaluation of choices except as they affect other choices. Its formal core valuation is simply that more good is better than less good.

Recent development of the theory of games goes a step further in generalizing the theory of rationality.[28] It works out the mathematical theory of "best" moves in games of different sorts. Insofar as social processes can embody the

---

26. For a survey of the present status of welfare economics and the problems in its various conceptions, see Kenneth E. Boulding, "Welfare Economics" in *A Survey of Contemporary Economics,* vol. II, ed. by B. F. Haley (published for the American Economic Association by Richard D. Irwin, Inc., Homewood, Illinois, 1952), pp. 1-34.

27. *Ibid.,* p. 30.

28. See, for example, Kenneth J. Arrow, "Mathematical Models in the Social Sciences," in *The Policy Sciences,* ed. by Daniel Lerner and Harold D. Lasswell (Stanford University Press, 1951), pp. 129-154.

model by conceiving of one playing against nature or against others, these formulations admit of application. Since the concept of "best move" means winning most, any application of the theory to a specific domain, such as economic behavior, has to interpret the goals quite concretely by specifying aims.

On the whole, then, <u>maximization conceptions of rationality employed to diminish ethical indeterminacy retain a hypothetical element, just as harmony concepts do.</u>, They refine modes of maximization in mathematical terms but leave indeterminate whether the assumptions required for their application can be legitimately invoked. These are assumptions of fact about men and their behavior and assumptions about specific lines of desirable policy. Accordingly, the major burden of diminishing ethical indeterminacy is transferred to these areas. Even the general accounting and calculating spirit of the maximization approach itself requires evaluation to see whether it misrepresents man's psychological nature, and how far its embodiment in human attitude and character is to be welcomed or rejected.

DISINTERESTEDNESS. The role of the question whether every interest is to be reckoned, in the theory of welfare, points to the special importance of disinterestedness as an element in the conception of rationality in ethics. Obviously, if it prevails it has a powerful influence in removing indeterminacy from ethical judgment. And in ethical theory it is sometimes included with a haste that does not pause for justification. For example, J. B. Pratt advocates a "principle of rationality and value" and identifies reasonable conduct as "unprejudiced conduct which is for the sake of the greatest value or good at issue."[29] It thereby enshrines disinterestedness simply by definition, so that the egoist who refuses to reckon with the communal judgment can then be dismissed as unreasonable: "*That* is the sort of fellow you are—limited, partial, self-blinded, irrational, excluding yourself deliberately from the company of reasonable be-

---

29. J. B. Pratt, *Reason in the Art of Living* (New York, Macmillan, 1949), p. 152.

ings, and deserving for the time being at any rate, to be classed among the unreasoning brutes."[30] Pratt's concept of reason rapidly expands until it even comes to be described as *love*.[31]

Inclusion of disinterestedness in the concept of rationality in ethics requires justification explicitly in terms of lessons of experience and proposed values. There is a long history of attempts to distinguish between a kind of personal point of view and a perspective which steps, even if only temporarily, outside of one's own narrow confines. It shows clearly that a number of different motivations underly the inclusion.

One motive springs from the recognition that a man may make mistakes. He must, in Aristotle's language, distinguish between the apparent good and the real good, the latter being what a man of practical wisdom would judge good. This obviously applies the lesson from the field of bodily health that something may seem good to a man who is ill but really be bad for him. Aristotle also carefully distinguishes practical wisdom from a mere cleverness or skilful cunning in getting what one happens to want.

Another factor in the theory of disinterestedness is the recognition that passion distorts judgment. Hence the British moralists from Butler to Adam Smith speak of reflection in a "cool hour" or of a man standing temporarily aloof as a kind of "disinterested spectator" within the individual. Butler carefully distinguishes even self-love from the passions. It is thus not a question of my passions versus other people's passions, but it may be my passions versus my self-love.

The crucial issue arises, however, in the relation of my *N.B.* self and others. Here disinterestedness means forgetting the particular self in reckoning what is right and wrong. Butler assumes too readily that self-love and conscience give identical verdicts. Kant's method of universalization, like the Golden Rule, takes you to another person's perspective to correct your own. Bentham's general assumption has every

30. *Ibid.*, p. 189.
31. *Ibid.*, p. 249.

man count as one, and Mill carefully expounds in his *Utilitarianism* the fact that you are supposed to decide in terms of the greatest happiness of the greatest number, counting yourself no more than any other person. Sidgwick, by far the most careful in his analysis, finds himself as a rational being bound to aim at good generally, and therefore to regard the good of another as much as his own. But he finally recognizes that a man is concerned with the quality of his existence as an individual in a fundamentally important sense in which he is not concerned with the quality of existence of other individuals, "and this being so, I do not see how it can be proved that this distinction is not to be taken as fundamental in determining the ultimate end of rational action for an individual."[32]

Our concern here is not with the whole theory of egoism but simply whether the concept of rationality in ethics should include disinterestedness. Insofar as disinterestedness means an over-all point of view capable of revising mistakes, insofar as it means discounting passion, it is clearly part of rationality, because the evidence on these points is adequate and the implicit values are accepted. Insofar as it means that other people have values of their own and any careful reckoning must include these as data, it also is generally acceptable—the most callous slaveholder wants to know in his "reasonableness" at what point the slaves will rebel. But should the disinterested element of rationality also include putting something aside in the reckoning for others' values? Should it involve seeing yourself as simply one in a group of many? Clearly such a decision would involve incorporating democratic equalitarianism into the concept of rationality. This is a policy decision which will have to be made on the basis of the best available evidence —psychological, social and historical.

It is important to note that the question of the degree of intrusion of the individual factor in an ethical decision is

---

32. Henry Sidgwick, *The Methods of Ethics* (6th ed., New York, Macmillan, 1901), p. 496.

not a single yes-or-no problem. Men may agree that in some fields the reference to themselves as individuals is irrelevant. Thus, the most selfish individual may allow that everyone should have only one vote, or that a tax budget of a certain amount is required to support public services, or that someone who can swim ought to rescue a drowning man. He may differ on how the tax load ought to be distributed, or whether *he* ought to be the rescuer. Simply because so many issues in modern complex life have a social character, it may be that some aspects of ethics can readily incorporate disinterestedness into the rationality concept. But this is already invoking the evidence of the present state of our culture.

Now even if disinterestedness is completely incorporated into rationality, the egoist may still reject it by rejecting rationality itself, either in a particular situation or in general. And he would be "rational" about it only in the more limited sense which excluded this value. At present we are probably in a transitional stage. In discussions of group welfare, the conception of rationality embodies disinterestedness. In discussions of individual welfare, this element disappears. This may be a reflection of special conditions in a culture which create wide gaps between social welfare and individual welfare.

GENERALITY. A further element in rationality, which ⑦ we already met in considering Kant's elaboration of consistency, as well as in the universalization implicit in disinterestedness, is *generalization.* The drive towards generality is clear in scientific work and the very conception of a science as a system of laws. But to what extent being reasonable in ethics entails the same conception and what tools it furnishes us for the forms that more determinate ethical judgment may take, are questions we have explored in considering moral laws. On the whole, the decision to incorporate the striving for generality into a conception of rationality in ethics involves in part a factual discovery that some forms of law may be serviceably employed, and in part the determination to seek the maximum of generality that the subject-matter admits of. Underlying such a resolution are the familiar aims of knowledge, control, security, which are the base of scientific endeavor in general.

SOME CONCLUSIONS. We have discussed the chief constituents of rationality in ethics as they appear in the historical treasure chest of that concept, the elements that philosophical theories of the past have on the whole stressed as most "constitutive." No doubt, if we dig deeper, we shall find other properties that stem from the special characteristics thinking has as a human activity. For example, thinking involves the use of symbols, and so rationality might be explored for its symbolic qualities, and perhaps in ethics these take a special form. Similarly, thinking has a probing character, a problem-oriented character, and perhaps this merits inclusion as much as the element of unity or system. But foregoing further exploration, what may we conclude about the appeal to reason in ethics and its role in reducing indeterminacy?

The appeal to reason has proved to be a complex conception. It combines methodological conceptions, which are sufficiently established to constitute defining elements, with assumptions about the psychological, social and historical operation of man, and with fundamental valuations in human life. The appeal to reason therefore carries weight only so far as these methods, factual assumptions, and basic valuations are established. Thus the methodological components rest on the successful growth of human knowledge, especially in the physical sciences, and the basic or core valuations this pursuit of knowledge entails.[33] Beyond that, one can proceed only with caution, and a step at a time. It is better policy not to broaden the conception of rationality too far at the present time, but to reveal as clearly as possible what elements it combines.

Where then does the individual stand in the ethical use of the appeal to reason? Let us summarize what it can do and what is beyond its present scope. To be reasonable you

---

33. See below pp. 261-262.

must be logically consistent—but this even allows you the(1) luxury of wanting opposite things at the same time. However, on the assumption that pain and excessive frustration(2) are evil, you should avoid conduct likely to produce breakdown or an inability to function—but whether you should worry about others' similar fate is another question. You should minimize conflict and pursue harmony and a value maximum—but where there are serious questions of general(3) direction you have to decide by the content, not by the de-(4) sire to harmonize or maximize. To achieve security and knowledge and control, you need some degree of generality(5) in ethical judgment—but don't press it farther than the subject will admit of, for that's unreasonable. Be disinter-(6) ested in the sense of recognizing the possibility of mistakes, and in avoiding the distortions of passion. Be disinterested in passing judgments of social welfare and in working out general moral rules—but keep an eye on how you individually will be affected. Yet, when you know all this, is it reasonable to give up some of your interests because of others' interests? Well, that depends . . .

Such a conception of the appeal to reason neither makes pretensious claims, nor on the other hand, abdicates wide scope and functions. It has many gaps, but does not of itself preclude the possibility that these will be filled by the *results* of scientific and historical investigation, yet it does not guarantee that this will happen. Thus it too points directly to the perspectives provided by the particular sciences.

III. Barriers and Tools: The Spectre of the Stubborn Man

THE PHENOMENON OF ULTIMATE DISAGREEMENT. Just as conscience, reason and the sense of moral law have served as central facts in the justification of ethical absolutism, so the phenomenon of ultimate disagreement has seemed to many to make an arbitrary relativism unavoidable. It is almost as if contemporary ethical theory were haunted by the spectre of the truly stubborn man. Thus Russell objects to Toulmin's account of ethics on the ground that it would not have convinced Hitler.[1]

The mode of thought which puts the stubborn man into his position of authority is familiar enough. As long as the argument concerns means, it is said, evidence for one or another view can be produced. But when you come to ultimate ends nothing can be done. Bentham advocates democracy, Nietzsche advotes the superman. You advocate equality, the Nazi advocates the supremacy of his group. As long as he is silly enough to advance his racial theory you can refute it by anthropological evidence. But if he simply says "The supremacy of my gang is what I stand for and what I am ready to die for" you are reduced to silence.

---

1. S. E. Toulmin, *The Place of Reason in Ethics* (Cambridge Univ. Press, 1950), p. 165.

You may fight him but you cannot refute him. You may discover the causes that made him that way—psychological, cultural, historical—but that does not prove him wrong. After all, there are causes that made you what you are today. As Bertrand Russell puts it: "Science can discuss the causes of desires, and the means for realizing them, but it cannot contain any genuinely ethical sentences, because it is concerned with what is true or false." And again, "Since no way can be even imagined for deciding a difference as to values, the conclusion is forced upon us that the difference is one of tastes, not one as to any objective truth."[2]

Ethical writings of the last few decades furnish numerous testimonials for the sense of ultimate disagreement. Here are some samples:

"Morality has as its basis, in other words, an arbitrary, underived commitment to certain of the possible guiding principles and purposes. Faced with an assortment of a number of possible selves, one must make a choice."[3]

". . . the realm of values is characterized by encounters between ultimate sets of postulates, of resolves, or of will-attitudes."[4]

"This is a battle fought between decisions, a struggle between wills—not a discrepancy of theories to be settled argumentatively, or an antagonism of values to be bridged valuationally. The decision between the opposite systems is a non-rational decision, a decision which is its own reason and which proceeds with the simple force of: I want myself rational; or I want myself racial!"[5]

"If I say 'What I want is good,' my neighbor will say, 'No, what *I* want.' "[6]

2. *Religion and Science* (New York, Holt, 1935), pp. 237, 238.

3. Charner M. Perry, "The Arbitrary as Basis for Rational Morality," *International Journal of Ethics*, XLIII (Jan. 1933), p. 138.

4. Donald C. Williams, "Ethics as Pure Postulate," *Philosophical Review*, XLII (July 1933), p. 402.

5. Walter Cerf, "Philosophy and This War," *Philosophy of Science*, IX (April 1942), p. 181.

6. Bertrand Russell, *op. cit.*, p. 231.

It is easy to see the vast array of conflicts in the modern world that support such a mode of thought—major clashes of interest, whether national, class, or individual. National conflicts are associated with the abandonment of reasoned argument and the ultimate appeal to war. Class conflicts break even the bonds of national unity and familial loyalty. Individual conflicts of interest gain an added ultimacy from the fact that a man has only one life to live, and that—unlike a class or nation—what he gives up is given up forever. Is it surprising that a modern man slips readily into the assumption that ethics must become silent in the face of ultimate disagreement or use some absolute basis?

The nature and degree of ultimacy in disagreement in morals and ethical theory is, of course, a central problem in this book. In this chapter, however, we are concerned only with the barriers which are thrown up against a fully scientific exploration of the problem by attempts to settle it in a semi-a priori fashion. And these attempts have been extraordinarily widespread and at times extremely subtle. With Protean versatility they have taken now methodological, now logical, now metaphysical, now linguistic form; and when these wholesale arguments are pinned down, there has been the attempt in retail fashion to pass the package of ultimate disagreement across the counter, but without allowing it to be opened for inspection. Some of these attempts may be dealt with briefly, others will require longer attention. But all provide instructive lessons for subsequent scientific inquiry.

§ Are Ultimate Premises Arbitrary?

A METHODOLOGICAL PROBLEM. All inquiry has premises. If we do not accept self-evident ultimate moral truths in the old rationalistic vein, then we must accept arbitrary stipulations as axioms. All moral conclusions are therefore hypothetical: if we reject the axioms we are not bound by the conclusions. A man cannot go on forever looking for the basis of the basis. "Not even a miracle," says Charner Perry, "could furnish a logical ending for the endless chain which would result from the attempt to find a logical beginning for his judgment."[7]

POSSIBILITY OF INDUCTIVE VERIFICATION. Such methodological argument establishes nothing special for ethical theory; it merely maps the structure of a pure mathematical or logical system. It explores neither the possibility nor probability of agreement and disagreement. And nothing in the argument prevents general principles in ethics from being treated inductively as general laws in physical science are. For here we are no longer dealing with pure mathematical system but with applied or interpreted systems. Thus, just as general laws in physics are responsible to the evidence that may be gathered for or against them, so ultimate values are responsible to the specific kind of life that will emerge in acting on them.

It is trivial to compare the choice between equalitarian ethics and racist or élitist ethics to the arbitrary taste preference between oysters and clams. The latter choice is indifferent precisely because the moral assumption is already established—that in matters of food, within certain limits, a normal healthy person has the right to follow his taste. This is a moral *conclusion*, not a first premise. If his choice instead is for roasted human brains our judgment would be quite different. Even the question who should have free choice in food may be empirically determined. For example, there have been experiments on whether a baby should be left to follow his own taste, whether over a month he will balance the proper diet through a sort of wisdom of the body.

Similarly, in the judgment that an equalitarian moral hypothesis is preferable to an élitist moral hypothesis, there is the whole field of history and psychology to provide evi-

---

7. Charner M. Perry, *op. cit.*, p. 139. This is in a section headed "Arbitrary Choice, or the Bottomless Pit of Reflection."

dence. Of course, there are value judgments in the evidence, since one does not simply test a value principle by conse- quences but by the *value* of the consequences. There are factual assumptions in every scientific experiment, that are not questioned in the experiment itself, but may be made the subject of a separate inquiry. In short, ethical inquiry may turn out to have the same type of non-vicious broad circular character that many philosophers of science ascribe to scientific inquiry—a character admitting of progressive self-correction in the light of growing knowledge rather than requiring arbitrary first principles. The differences between the ethical model and the physical model may be profound. But they cannot be established by an argument from pure methodology.

On the other hand, if it be claimed that ethics requires ultimates in some sense that science does not, or that veri- fication of value hypotheses is only verbally similar to veri- fication of scientific hypotheses, or that science has precise operational standards and moral inquiry does not, then we are launched on a careful detailed comparison between the physical, psychological, social and historical sciences, and the field of morality. This requires investigation of the ma- terials of the field in all its detail, which is precisely the avenue we propose against the semi-apriorism of the arbi- trary relativist.

§ The "Unbridgeable Chasm" between Fact and Value

ON GOING FROM THE "IS" TO THE "OUGHT". Any attempt to treat morality as if it makes assertions capable of verifica- tion is immediately greeted in contemporary ethical theory with the outcry that it rests on a logical confusion. Ethics, it is said, is not a kind of psychology. Psychology tells us what *is* the case, and ethics what *ought to be* the case. A judgment of value cannot be derived from a judgment of fact. From a statement about what *is* the case, we cannot logically conclude a statement about what *ought to be* the case. Once again the implication of this argument—or rather its "and therefore"—is that the value judgment is scientifically irresponsible and has either non-natural controls or no controls. The arbitrary relativist thinks in terms of the latter.

MUCH ADO ABOUT LITTLE. That you cannot draw a value conclusion from factual premises has become such a commonplace in the claim for the non-scientific character of morality, and has been so often enunciated pompously by eminent philosophers and peremptorily by social scientists arguing for the value-free character of their field, that I need offer here no samples or illustrations. But I cannot repress a sense of amazement at the proportions to which a limited logical proposition has been blown up, and the weight of theory that it has been made to bear. It is true, roughly speaking, that a categorical assertion containing a given term does not follow validly from premises which do not contain that term. No conclusions about what ought to be or what is good can be drawn from premises solely about what is. No conclusions about colors can be drawn from premises solely about light-waves; you have to introduce premises relating colors and light-waves. No conclusion that Socrates is mortal can be drawn simply from the premise that all men are mortal; you have to add that Socrates is a man, thus introducing the term "Socrates" into the premises. All the laws and observations of physics and astronomy could not serve to deduce that an eclipse will take place unless some premise defining the term "eclipse" is introduced or some rule for its use provided. This warrants no conclusion that an eclipse is a non-physical phenomenon with overtones of suspicion that it may be an instrument of Zeus or else a human fiction! Similarly we cannot on this logical ground alone proclaim the uniqueness of Socrates, or the mental as against physical status of colors. And so nothing is established likewise about the independence of value and fact. In each case the underlying problem of dependence or independence is a question to be settled by examining the phenomena, or else it is an issue concerning what kind of definition of the term or rule for its use is to be introduced and what grounds for its acceptance or rejection are to be permitted.

It may be argued that limited as the logical point is, many absolutisms have rested on ignoring the logical rule. It is said, for example, that God willed no stealing and therefore stealing is wrong. Or evolution is moving towards something, and therefore that something is right. But such formulations may be viewed as *enthymemes* in which one of the premises is suppressed. Half the use of logic is in trying to find out what the suppressed premises in discussion are. An able partisan of these ethical positions will know what underlying definitions of moral terms or underlying theoretical assumptions he is making. An uncritical novice should be helped to discover his assumptions and definitions. No ethical view is crushed by hurling the logical maxim as a thunderbolt.

THE IMPORTANT UNDERLYING PROBLEM: DEFINING ETHICAL TERMS. The real problem generated by this whole barrier is the underlying issue of defining ethical terms. And here there is sufficient difficulty to make us pause. In fact the history of ethical theory shows that the greater part of theoretical controversy is often cast as a conflict of such definitions. A full treatment of the problem requires a comprehensive treatise. And yet without some treatment, issues of this sort will haunt us in all the specific analyses that are to follow.

The fundamental difficulties in definition are not peculiar to ethics. Comparable problems are found in the history of such terms as "matter" and "force" in physics, "intelligence" and "personality" in psychology, "culture," "state," "democracy," "value," "institution," and so on, in the social sciences. The fact is that an adequate definition reflects an established theory of the field, and this involves extended knowledge on the one hand, and agreed-on purposes in the cultivation of the field on the other. On the whole, then, it is advisable to suit the type of definition employed to the stage of development of the subject and not to give the appearance of

completion, closing off further inquiry into the justifying
theory underlying the definition.

MODES OF DEFINITION.   Contemporary development of
the theory of definition has provided a whole array of modes
in relation to specific purposes and stages of inquiry.[8] Some
modes are useful as starting-points. Thus some investigators
start by *stipulation* from the area in which people using the
English language apply the terms "right," "wrong," "good,"
"evil." The stipulative definition serves to delimit the rough
original field to be regarded as morality. Any further developments, including more careful delimitation of the field
itself, come from the results of subsequent analysis. Other
investigators start from examples, such as that one ought
to keep a promise, or that it is wrong to kill a human being.
Such *ostensive* or *denotative* definitions likewise serve as
starting-points.

Operational definitions are more often suggested, used
and refined in the course of the investigation. For example,
we saw above[9] that special feelings may serve as pointer-
readings for obligation terms; similarly, being pleasant or
being the object of desire has often been used as a test for
"good." Operational definitions and kindred types constituting a kind of partial definition are common in scientific
work.[10]

The ultimate aim is a definition which would state the
necessary and sufficient conditions for the application of the
fundamental terms in the field under investigation. (These
have at different times been called *real, structural* or *theoretical* definitions.) Such definitions do not just come to
hand lightly. They are the outcome of a long development
of knowledge, and it is unwise to forget the relationship of
the definition to the underlying knowledge claims which

---

8. Cf. Carl G. Hempel, *Fundamentals of Concept Formation in Empirical Science,* International Encyclopedia of Unified Science, vol. II,
No. 7 (Chicago, The University of Chicago Press, 1952), pp. 2-50.
   9. p. 40 f.
   10. Hempel, *ibid.,* pp. 23-29.

would justify it. Where there is not enough knowledge, several candidates may compete for the post, and the tenure of any occupant is tentative pending further investigation. For example, an "ultimate" definition of "intelligence" in psychology is unavailable because there is not yet sufficient knowledge of the terms that will enter into a theory of the nature of thinking adequate to the phenomena involved. An "ultimate" definition of "culture" is likewise unavailable because we do not as yet have a sufficiently extensive picture of the dynamics and interrelations of a society to provide a unified theory of its operation.[11]

An "ultimate" definition of "good" and "ought" depends in similar fashion on getting a "complete" picture of the nature of man, his cognitive faculties and social relations. As it is, we do not yet have agreement on which phenomena are to be regarded as primary in ethics. There are two major classes. One is the family of what we may call *aspiration* phenomena, conceptualized as the good, the other *obligation* phenomena, sometimes seen in individual terms of conscience and guilt-feeling, sometimes in interpersonal terms as mutual claims, rights and duties, but all sharing a certain "binding" quality. Different ethical definitions often reflect differing interpretations of the relation of these phenomena, some giving a primary role to "good," others to "ought." For example, the hedonists, resting on the hasty psychological assumption that pleasure is the ultimate goal of all human activity, equated good with pleasant and brought all the other terms into subsidiary relationship. Still other views expect no ultimate unification but rather a permanent duality or plurality of concepts and phenomena.

The postponement of "ultimate" definitions need not, however, mean that ethical terms are inherently indefinable or anything of that sort. They may simply be regarded as open concepts, as many scientific concepts are. Or else it

---

11. See the numerous definitions and the many perspectives underlying them in A. L. Kroeber and Clyde Kluckhohn, *Culture, A Critical Review of Concepts and Definitions* (Papers of the Peabody Museum, Harvard University, vol. XLVII, 1952.)

may mean that several candidates are being considered, resting on competing theories of man's nature. For it seems more important—now that the theory of definition has relaxed somewhat the previous philosophical tension about such problems—to spotlight the underlying theory and to choose one's mode of defining deliberately than to argue definability versus indefinability or to rush into premature definition of a perfectionist type.

A WORKING CONCEPTION. Still, for our present inquiry, if we are to talk of *good* and *evil* we need some working conception of their meaning. There is, fortunately, some minimal linguistic agreement among most ethical theories. They tend to define "good" to some extent in terms of the direction of human striving or aspiration, and "evil" as the object of aversion or at least what profoundly frustrates the good. This is enough for our purposes, and it achieves a comparative neutrality. It enshrines neither a naturalism nor a supernaturalism, neither a collectivism nor an individualism, neither a subjectivism nor a realism. Such differences would be capable of formulation *within* the common linguistic agreement. Thus some would find striving to be the expression of impulse, others the more or less conscious direction of the self to a transcendent goal. Some would find a common object for striving, others only individual variable goals. Some would analyze striving as personal feeling, others as discernment of objectively real qualities. Nothing is therefore begged in advance by the usage we propose to follow. And if we need further qualification, or if we need a comparable stipulation for obligation terms, they can be introduced at a subsequent point in the specific context.

§ Ultimate Commitment and the Causal Principle

THE APPEAL TO CAUSATION. Another wholesale argument to bolster the sense of ultimate disagreement—this time metaphysical rather than logical—is the appeal to a belief in causation. Surely, it is said, we grant that every-

thing that exists has a cause. Then a man's values too have
causes. My ultimate commitments are what they are because
of my own peculiar physical, psychological, cultural, historical conditions. Had these been different they too would
have been different. There is no sense in talking of justifying
them. You may try to move me by reconditioning processes,
seductive emotions, even threats. But this is simply what
you are at bottom versus what I am at bottom. Think of
Luther defending his position ultimately with the proclama-
tion: "Here stand I; I cannot otherwise." It is true that he
appeals to God immediately thereafter, but that is a separable issue from the phenomenon of ultimate commitment
that he illustrates so simply. The last word is the *bare* fact
that individuals commit themselves to values, and this ends
justification and leaves room solely for causal explanation.

How Bare Is Commitment? The argument just advanced is very powerful in contemporary, especially naturalistic theory, and requires careful examination. If we
unravel its several strands, we can see that it does not
answer the question of ethical relativity; it merely *poses* it.

In the first place, to see how *bare* the fact of commitment to values is requires itself a full scientific investigation. It may prove to be as richly clad as motion is in physics,
as growth in biology, as striving or goal-seeking is in psychology, and as mass-movements or national aspirations are
in history. Even a cursory examination shows people do not
take on value-commitments haphazardly. A phenomenology
of value-acquisition is a complex inquiry: some values grow
up wild, others are cultivated; some hook on to what is there
already, some are a kind of mutation appearing in the field;
some cling inside the self, some on the edge of the self;
some parade outside to show their finery. It is a priori
metaphysics, not appeal to an obvious fact to argue that
there *must* be an ultimate commitment to values that is
contingent, as if a naked self just simply chose. It seems
more likely that men are never without some values, that
the contingency of the commitment is never that of a bare
self asking for value clothing. It is always the same kind of

contingency that sense-perception has in science. It is simply the fact of specific desires, needs, aspirations of particular men at particular times which is the matrix in which evaluation occurs. Ultimate disagreement of interests may and does occur as a descriptive matter, although what precisely the test is for the fact that it is ultimate, is a quite neglected question.

But even where it does occur, ultimate disagreement is not necessarily a uniform phenomenon. Luther's "I cannot otherwise" is not of the same type as a case of arbitrary taste, irresistible provocation, or kleptomania. Nor is the role that ultimate disagreement of interests is to play in ethical theory settled by the fact that it exists, without considering the aims embodied in ethical theorizing itself.

THE LUTHER-PHENOMENON. The fact of principled ultimate commitment is itself so significant for ethical theory that it deserves a special name. Let me christen it as the *Here-stand-I-I-cannot-otherwise-phenomenon,* or since this is cumbersome, the *Luther-phenomenon.* And a brief examination will show that it is a rich phenomenon already embodying moral claims and causal elements.

What requires explanation is not merely what Luther was committed to—in terms of the history of the growth of Protestantism, Luther's personality and social milieu, etc.,—but also the startling phenomenon that he includes his inability to do otherwise in his justification. Why should not this be taken rather as a mark of perverseness, of hardened heresy, of a corrupted spirit? The fact is that it was so taken by the opposition, in line with the familiar treatment of heresy as a kind of inner corruption shutting out the light. To treat the inability to do otherwise as part of the justification thus requires a special theory of its nature. This may be theological, such as the belief that each individual is capable of receiving the light without intervention of religious organizations; or it may be a moral view itself, a kind of individualism which says that an individual's ultimate convictions deserve respect.

IMPORTANCE OF DISTINGUISHING DESCRIPTIVE, CAUSAL

Very good

AND MORAL ENTERPRISES. To show that two men take ultimate opposite stands is description. To add that neither can
do otherwise is either adding their introspective reports—
which are data for further analysis—or else asserting a causal
account of the determination of their outlooks. To regard
the ultimacy as the end of justification is again descriptive:
it repeats the fact that these commitments play a specific
"ultimate" role in the justification-processes of each of the
individuals. But it provides no basis for any conclusion whatsoever in the comparison of the merits of their stand. This
would be a moral judgment within a moral framework. One
cannot then slip into such familiar formulations as "Each is
right from his own point of view" or "No man's values are
superior to any other man's values," unless there is a definite
moral assumption of some kind. For example, if it is said
that each man is sincere, or that we have no right to inflict
our point of view on anyone else, and so on, it is assumed
that sincerity or tolerance or mutual respect or human equality or individualism are major virtues and values. In these
cases ultimate differences are being justified by a moral liberalism, not solely by the fact that they occur as ultimate.

Finally, if the appeal is simply to the causal principle
and the consequence that sharply different causal histories
may produce sharply differing value systems, this is a proposition of the natural history of mankind. But it settles nothing about how different actual value systems have in fact
been, whether there are limits to variation, whether in the
midst of the greatest variety there may not be some common
bases below the surface adequate for a large measure of
comparative evaluation, and so forth. The causal principle
by itself, even when referred to ultimate commitments, cannot therefore bear the weight of an a priori relativism.

§ Language as Servant, not Master

GROWING IMPORTANCE OF LINGUISTIC ARGUMENT IN ETHI-
CAL THEORY. A great deal has been written in recent years
about the relation of ethics to language. We noted in chapter

I that a linguistic thread is found in the design of ethical relativity; and the emotive theory of ethics, itself fashioned through linguistic analysis, was cited as the most relative of relativisms. Sometimes it almost seems in ethical theory to-day that language is master not servant, that an examination of the linguistic use of ethical terms will of its own lay down the law concerning the nature and function of ethical theory.

Until recently many such linguistic analyses insisted that moral judgments were really imperatives not indicatives in the way they were used, so that to say that something was wrong "really" said "Don't do it." And this, of course, had the consequence that moral judgments could be neither true nor false. It is true that such simplifications were corrected by further linguistic analysis. It was seen that the functions of moral judgment in various contexts were many, that a fuller array includes expressing emotion, ordering, persuading, commending, ascribing rights and responsibilities, ceremonial utterance, utterance performing or creating relationships, goading and guiding. And the question could even be raised whether in all this array of possibilities one was not overlooking the function of teaching or, from the hearer's point of view, learning, and therefore the cognitive content of the moral judgment might once again recover its central place.[12]

_N. B._

---

12. I have reviewed the development of the linguistic trend and its forms, suggesting this outcome, in a paper on "Ethical Reasoning," in *Academic Freedom, Logic and Religion,* ed. Morton White (Philadelphia, University of Pennsylvania Press, 1953). For the various types of analyses, the expressive is found in A. J. Ayer, *Language, Truth and Logic* (London, Gollancz, 1936), ch. 6; the emotive in C. S. Stevenson, *Ethics and Language* (New Haven, Yale University Press, 1944); the commending in R. M. Hare, *The Language of Morals* (Oxford, Clarendon Press, 1952); ascribing rights and responsibilities in H. L. A. Hart, "The Ascription of Responsibility and Rights," *Proceedings of the Aristotelian Society* 1948-49; ceremonial utterance in Margaret Macdonald, "Ethics and the Ceremonial Use of Language," in *Philosophical Analysis,* ed. by Max Black (Ithaca, Cornell University Press, 1950), pp. 211-229; performatory utterance in J. L. Austin, "Other Minds," *Proceedings of the Aristotelian Society,* Supplementary vol. XX, esp. pp. 169-174; goading and guiding in W. D. Falk, "Goading and Guiding," *Mind,* LXII, No. 246 (April 1953).

cf. C. Kluckhohn, Mirror for Man, ch. 6, "Gift of Tongues", pp. 128 ff.

More important than decision among these specific functions is the whole question of the relation of language analysis and ethical theory. To take a physical parallel, Whorf points out that in the Hopi language the universe can be described without using a concept of dimensional time. A physics along these lines would therefore work without its time and velocity concepts. He thinks that this could be accomplished "though of course it would require different ideology and perhaps different mathematics."[13] Now what would we think of a hypothetical Hopi treatise which while Hopi scientists were struggling to develop the requisite mathematics, sought to show statements in which the term "velocity" appeared were meaningless, and that there could accordingly be no science of mechanics? My point is simply that important as the study of language and the analysis of linguistic uses undoubtedly is, in the last analysis language is servant not master, and men's conceptions of fact and men's actual valuations call the tune. Now it is true that language forms may hinder the grasp of certain types of fact or enshrine certain types of valuation. Linguistic analysis may therefore perform the invaluable service of revealing and removing such limitations; it should not reveal in order to enshrine.

The latter is precisely what the emotive theory seems to me to have done for arbitrary relativism. It fashions an ethical language in such a way as to reflect the belief that ethical judgments are incapable of scientific verification and that their core is the expression of emotions. Thus Stevenson puts the phenomena of ultimate disagreement squarely before us at the outset and then weaves a language to give them a central place in ethics. Ethical disagreement is found to involve "an opposition, sometimes tentative and gentle, sometimes strong, which is not of beliefs, but rather of attitudes—that is to say, an opposition of purposes, aspirations, wants, preferences, desires, and so on."[14] Moral judgments

13. B. L. Whorf, "Science and Linguistics," reprinted in Newcomb and Hartley, *Readings in Social Psychology* (New York, Holt, 1947), p. 217.
14. C. L. Stevenson, *Ethics and Language,* p. 3.

are found to be engaged in *recommending* for approval or disapproval.[15] This element of expression or persuasion, this mutual antagonistic pressure, is assigned a dominant role. By incorporating this emotive element into the meaning of ethical terms, Stevenson fashions the models he regards as most serviceable:

"(1) 'This is wrong' means *I disapprove of this; do so as well.*
(2) 'He ought to do this' means *I disapprove of his leaving this undone; do so as well.*
(3) 'This is good' means *I approve of this; do so as well.*"[16]

Now in such a language it seems impossible to prove that one side in an ethical disagreement is "wrong" and the other "right," as one would with two conflicting propositions in science. One might offer evidence for beliefs that the parties hold, and hope that they may shift their attitudes. But, as Stevenson sees it, it is an intellectualist fallacy to think that the mere disproof of beliefs will always have this effect, that all disagreement in attitude can be reconciled by reconciling disagreement in belief. "It may always be," he says by way of example, "that when nations dispute a colonial right, all but one would withdraw their claims *if* the full truth were known about consequences, precedents, motives, and so on."[17] But he suspects such an assurance, and no doubt with good historical grounds. But note the unquestioned assumption that *if* the competing claims were not withdrawn there is no way of judging which is right or whether all are wrong. All we really have is the variety of mutually antagonistic pressures. And if an agreement is reached, all the participants could presumably say, "That's good," meaning no more than to indicate their approving attitude and encouragement of others to do likewise.

The mode of speech, once it is established, thus gets one out of the habit of asking which party to a disagreement is

---

15. *Ibid.,* pp. 12-13.
16. *Ibid.,* p. 21.
17. *Ibid.,* p. 137.

"really" right. In fact, the only meaning it seems to leave for a *solution* is not the scientific one of an *answer* to the problem but a practical one of *actually getting the parties to agree*. If we cannot, there is ultimate disagreement.

CRITIQUE OF THE EMOTIVIST FORMULATION. The first temptation may be to fight against the language. One may ask why we should use Stevenson's models, what lessons of experience and policy are involved in adopting them as compared to Bentham's or Dewey's or Aristotle's or Aquinas' or anybody else's. Such an inquiry is indeed an interesting one: it calls for a greater formalization of the language and a greater explicitness in presenting assumptions of fact and value on which the preference of one over another is based. But let us note that *where scientific results are adequate* a way can usually be found to break through the barriers of any language. For example, the psychiatrist will not be stumped by Mr. A's ultimate disagreement of attitude if he happens to know that Mr. A is neurotic and that only a successful analysis would shift his attitude. As a matter of fact, if Mr. A were incurable, he might never change his attitude. Now how can the psychiatrist give an empirical meaning to the judgment that A is wrong? He might remind us that the psychoanalytic process involves—almost in Socratic fashion—a man coming to know the character of his own present attitudes, hence gaining an insight he did not possess and which he was at the time of the argument incapable of having. We must not over-simplify the problems actually involved in whether the analytic situation is primarily a growth of insight or a process of causal alteration; this would require detailed consideration. But it does involve in some sense a growth of self-knowledge. Hence the psychoanalyst's judgment of Mr. A's ultimate attitude may be formulated something like this: "If A knew facts of such-and-such a type, *which he is at present incapable of knowing,* he would abandon his attitude, and, perhaps, adopt B's." Now since he may never be capable of knowing them, then how long B continues to dispute with him and whether A in fact changes his attitude, is irrelevant. The formulation in terms of the contrary-to-fact conditional provides a meaning for saying that Mr. A is *wrong* without making it necessary to wait for his lordly agreement. The veto power of a present ultimate attitude is gone in this particular case. In its place is a responsibility to scientific knowledge of human life—psychological, cultural, social and historical.

It is important to note what this establishes and even more what it does not. It does not set up the psychoanalyst as an authoritarian father figure. It does not say that we can psychoanalyze colonial powers or take all conflicts in our stride; there are certainly descriptively ultimate conflicts, and some of them *may* be moral, but that is a separate issue. All it does show is that in one small area ultimate disagreement need not stop a further judgment of who is correct. We have not here, then, a restored moral absolutism based on esoteric experts, but an appeal to scientific knowledge for analyzing the particular case. The issue is not to overthrow ethical relativity, but to break the barrier of a semi-a priorism that rests it on the mere fact of ultimate disagreement of attitude. How far this mode of argument can be extended—for example, whether whole sets of attitudes can be labelled "rationalizations" or "ideologies" and therefore perhaps judged wrong in spite of the sincerity with which they are held—cannot be settled by a priori conjuring. Whether it can be done or not depends on the kind of factual evidence available and the kind of scientific theory that we may find established in human affairs. What has really been accomplished, then, is to show that the fundamental issues we are discussing are not linguistic ones, and that no such screen should be placed between us and the actual exploration of the phenomena of ultimate disagreement.

§ Towards the Scientific Study
of Ultimate Disagreement

WHAT IS ETHICAL DISAGREEMENT? Not every disagreement in attitude that struggles for persuasion turns out to be an ethical disagreement. One may grant the reality and

intensity of conflicts of individual, class and national interests, if by that we mean desires or specific aims descriptively
considered. Obviously the two men on the log at sea cannot
both live if the log will hold only one, profits go down if
wages are increased, and the same oil of Iran cannot flow
in two directions at the same time. But sometimes, as very
often has been the case in the history of war, conflict occurs
in the context of ethical agreement—for example, where both
parties take the view that to the victor belong the spoils.
In fact here the ethical principle of distribution according
to strength or chance in victory itself helps precipitate conflict. An ethical evaluation of such a situation involves a
critique of the let-them-fight-it-out principle; no solution
is reached by giving each forceful claimant some fresh area
of his own to plunder.

Sometimes again, conflict of desires may exist and issue
in strife even where one side recognizes that the other is
right and that it is itself yielding either to temptation or
expediency. Of course, it may try to persuade itself that it
has some justification, but in sober moments this falls away.

Perhaps the simplest mode of identifying an ethical disagreement in attitude is still the familiar one—each side
thinks it is *right* and the other side is *wrong*. Nor is this
purely a question of words. Opponents with a spoils attitude, although engaged in war, may avoid bombing each
other's factories, or even seek to engage in indirect trade
during the war; whereas a scorched earth policy on one's
own land is more indicative of a principled conflict. It is
not conclusive since there may be a calculation of interest
involved, or the simple all-out effort against extinction.
But apart from judgment in terms of behavior, the mark of
ethical disagreement seems to be that there is some conception or other, cast in an ethical framework on the part
of the participants, however that framework may be analyzed, and that by a combination of behavioral and phenomenal investigation it can be more or less identified—at the
very least by distinction from the spoils attitude or the
yielding-to-temptation situation.

NEED FOR SCIENTIFIC EXPLORATION. In general, the phenomenon of ultimate disagreement requires psychological and historical exploration far beyond that found in the arbitrary relativism which enshrines it as a central fact of ethics. At the very outset, more explicit criteria are required for distinguishing ultimacy in a given disagreement. The arbitrary relativist has tended to operate with an unanalyzed concept of the self as the agent in disagreement and thus to treat its reaction as a simple act of will and self-assertion. Therefore he thinks only in terms of abstract commitment, missing the rich content of the real phenomenon, and he sees all solutions in interpersonal relations as simply agreement of will-attitudes. It is not surprising, if differences of taste, of interest, of principle, of judgment in human affairs, be cast in the single mold of will-assertion (and that for groups as well as individuals), that ethics can be described only in terms of attempted mutual persuasion in conflict situations, or else an agreement to disagree.

THE PLACE OF ULTIMATE DISAGREEMENT IN ETHICS. How ethics can best be described, and whether the phenomenon of ultimate disagreement should have a central, intermediate or peripheral role, is a question best faced in the light of the *results* of the fuller psychological and historical exploration we have found necessary. A priori, one could give it the central role that an arbitrary relativism has given it. A priori we could also give it a minimal place: for example, it is possible that men feel they have an *ethical* disagreement only in contexts already containing some implicit common standard. Ethics could then be defined as the effort of men who have some basic agreements to widen the area of common agreement. A priori, there are also intermediate roles to be given to the phenomenon of disagreement that lie between these extremes.

Other areas of life have faced decisions comparable to such delimitations of ethics. Medicine, once construed as simply curing the sick, has come to be thought of, with the advent of preventive medicine, industrial medicine, public health, as the whole effort to safeguard and maintain the conditions of health of the community. Law, so often construed in legal philosophies as simply a mode of settling disputes, has moved in a wider perspective into an instrument of social control and to some degree social engineering. Certainly one would not want to consider auto-driving with all its joys and opportunities simply in terms of avoiding collisions! The conception of ethics and the role that disagreement may have in it may be undergoing similar transformation at the present time, in the light of the growth of knowledge and emerging patterns of valuation.

POSSIBLE MODES OF DEALING WITH ULTIMATE DISAGREEMENT.   Actually, the possible methods of exorcising the spectre of the stubborn man are pretty standardized, in fact quite old-fashioned. There are really, after all, only three ways in which to overcome him, and these can be seen in the theological tradition quite as clearly as in the materialistic theories. The stubborn man learns God's will, and then says, "Why should *I* obey it?" The first answer is a threat— from Zeus' thunderbolt to the Christian Hell. The second answer is the appeal to sympathetic response: God enshrines justice and you want justice, therefore you are on His side or He is on yours. The first two both accept the YOU and God's will is justified in your terms. The third begins to shake the YOU. It says in a tone that varies from wrath to scorn, "And WHO are YOU?" If you are of a trusting disposition you may surrender at the first forceful query, as Job did, and become resigned. But if you insist on tracing all the connections the theologian will take you to pieces, show you that you are a child of God and that everything you are doing is really a disguised way of getting home again. The same holds for Hegel and the self-realization theories, according to which the will of the community is your real will. There is nothing wrong in general method with these approaches. The only question is whether they present an accurate account. But the obviously propagandist nature of many such conceptions that tie the individual to the existing institutions of the community has only added to the intensity of the stubborn man's obstinacy.

And he is wise in refusing to accept a scientific orthodoxy to replace a religious or an idealist one. His critical spirit must not be abandoned when his stubbornness is cast aside.

CONCLUSION. How far ultimate disagreement may in fact be dissolved, or whether it remains central in ethical theory, are thus questions requiring extended scientific exploration and policy decision based on the *results* of such investigation. Just as ethical absolutism could not be established in semi-a priori fashion by the feelings of conscience, the properties of reason or the sense of moral law, so ethical relativity cannot be established in semi-a priori fashion by the phenomenon of ultimate disagreement, real as conflicts are among human beings, nor by the many devices that have sought to give disagreement a central and dominating place. In analyzing these devices we saw that "ultimate" premises admit of inductive establishment, that satisfactory definitions are not created by fiat but grow with increased knowledge of the material, and we formulated a working conception of "good" in terms of the direction of striving or aspiration to serve as a starting-point. We saw also that in order to avoid confusion, a careful distinction must be preserved between phases of investigation, such as the descriptive, causal, and moral or evaluative. And finally, we sought to show the ultimately secondary role of linguistic formulations in ethical theory, and to break the ground for a scientific treatment of the phenomena of ethical disagreement.

IV. The Nature
and Sources
of Indeterminacy

Where do we stand in our philosophical analysis of ethical relativity?

We saw that the heart of the matter was the problem of indeterminacy—the question whether definite answers are attainable in moral judgment and ethical theory.

We saw that traditional ethical theories all had dependency-points at which the problem of indeterminacy had to be faced, that it could be obscured but not eliminated.

We saw that the ready answers given in the absolutist tradition and in the relativist tradition did not settle the problem. There are "unknowns" about man's feelings and situation, his world and his methods of acquiring knowledge, his interpersonal and social relations, that require a wider knowledge, which in our time science is trying to furnish. The analysis of the problems gave us valuable conceptual tools, but answers in ethical theory require a philosophical utilization of scientific materials,

We have now to adopt a more definite attitude to the indeterminacy problem. Certainly there is no reason to be afraid of indeterminacy or to quake in its presence. We need to understand the terms of our general relation to it and to be ready to look it in the face. To do this we need a

92

view of its structure and sources in generalized terms to guide our formulation of the search for the contributions of the various sciences. And we can get some help in fashioning a strategy for coping with indeterminacy by examining models from other fields.

§ Why Seek to Diminish Indeterminacy?

WHY INDETERMINACY ITSELF CONSTITUTES A PROBLEM. To propose, as we are doing, that we consider how to cope with indeterminacy is in some sense to take up an "anti"-attitude towards it, to see it itself as constituting a problem to be solved. That it does constitute a problem for man will probably be granted readily, but it is well to be explicit. For a priori, one might ask why diminish indeterminacy, why not leave it alone, or even increase it?

In a general sense, to ask questions and to face problems already involves some demand for an answer or a solution. To this extent those who ask the questions or face the problems are already seeking to diminish indeterminacy. They do not necessarily seek its complete elimination, since some indeterminacy within the answer is often permissible; for example, we are satisfied with a predicted temperature range for tomorrow rather than an exact mark. But if I ask the doctor "Will I get well?" and he answers "In this kind of case you either recover completely or else die" the indeterminacy strikes deep. In moral questions especially the indeterminacy is disturbing. One wants to know whether proposed ends are good, what obligations we have toward others, which character-traits are worth developing or strengthening. We can take the answer "It depends . . ." up to a point. For if we know what it depends on, we have somewhere to go and look. "Should I jump into the water to rescue a drowning man if it involves a risk of my life?" "It depends on whether you can swim." That kind of dependence is readily disposed of. "But it also depends on how high you value your own life." "I suppose I am as selfish as the next man, but should I be in such a case?" "It depends on what your goals in life are." "But what goals are good for me to strive for?" "It depends on what you mean by 'good' and what your desires and capacities are." "Well, here is an inventory of the latter, but what should I mean by 'good?' " "Thanks for the inventory, but that's not enough by itself. As for the meaning—it depends what school of thought you belong to." "But what school of thought is correct?" "What makes you assume there is a correct answer?"

It is easy to caricature the relativist. He seems to offer an endless chain of indeterminacy. And if we choose the example of the drowning man, we feel a sense of discomfort as the discussion drags on. We should be up and doing; either jumping in or getting help. To be fair to the relativist we should choose the complicated cases—statesmen faced with decisions of war or peace, men choosing between principles and jobs that support their families, conflicts of national interests, class interests, personal loyalties, and hosts of prevailing issues that press for decision and cannot be avoided. In many such cases, even after a choice is acted on, men will want to know whether they have been correct in their judgment. In the medical example, I will know eventually whether I am recovered or dying. In a moral problem will I ever really know that I have done right, if there is an endless sequence of dependency? Is there any wonder that men rush to the absolutist for comfort as well as for guidance?

INDETERMINACY SOMETIMES DESIRED. On the other hand, it is also easy to caricature the determinateness of the absolutist, just as we did the indeterminacy of the relativist. It is part of the stock in trade of ethics texts to show that simply sticking to principle without paying attention to facts is to court smugness and disaster. We saw that Kant argued that one must not tell a lie even to a would-be murderer who asked where the intended victim had gone;[1] on which Hastings Rashdall comments that under English law, if he acted in this way, Kant would probably be con-

---

1. See above, p. 43 f.

sidered an accessory before the fact![2] Adam Smith lists
moral philosophers who would say you should keep your
word to a highwayman who forced a promise from you to
pay a sum of money, even though circumstances have so
changed that he can no longer put pressure upon you. He
himself proposes a compromise instead: if the sum is only
five pounds, pay it, but not if it is so large as to ruin your
family![3]

Certainly not all ethical determinateness at any price is
to be welcomed. Sometimes the removal of indeterminacy
stems from a great evil, such as a flood or a sinking ship,
which makes the moral issue simple and determinate by its
urgency. Where all hands are needed at the pump, alternative courses of decision may quickly become morally inap-
propriate. In particular cases we may also decide to maintain
a special form of indeterminacy—for example, the decision
not to apply an established rule automatically, a decision
on a particular occasion not to pursue further knowledge, a
decision not to decide but to let events take their undetermined course, and so on. Sometimes we actually want to
widen areas of indeterminacy; for example, in raising the
standard of living and so increasing alternatives of choice,
we convert the very possibility of varying selection into a
human good. But while this renders individual decision less
determinate, it involves a very determinate moral framework in the principle of freedom of choice and the right to
exercise taste. And in all the other particular examples of
deliberate indeterminacies the deliberateness is justified
within a definite framework. The result is, as it were, not
simple indeterminacy, but bound indeterminacy, harnessed
to human aims.

MINIMIZING INDETERMINACY AS A GENERAL POLICY.
There is thus no simple antecedent justification of a general
policy to minimize indeterminacy in ethical judgment. It
rests upon the conviction that knowledge and deliberateness

2. Hastings Rashdall, *Ethics* (New York, Dodge Publishing Co.), p. 57.
3. Adam Smith, *The Theory of the Moral Sentiments* (London, G. Bell
and Sons, 1911), p. 486.

have a useful role to play in human life, upon the recognition that there are human problems many and urgent, and that this is a perennial rather than a temporary feature of human life. And in our contemporary setting it rests on the pressing nature of the issues which have focussed attention on the demand for some bases for a common-human morality. But the fullest justification in the long run can be provided only by the results of our inquiry, not by initial stipulation.

§ Structure and Sources of Indeterminacy

STRUCTURAL FACTORS IN THE PROBLEM FIELD. In formulating a generalized conceptual scheme for treating the problem of indeterminacy we may take as a starting point John Dewey's use of the idea of a problem-situation in ethical theory.[4]

The essence of Dewey's approach lies, as I see it, in what is really a general scientific hypothesis or heuristic principle. It is the insistence that ends or goals as they arise in human life do not grow at random. They express problem-situations in which individuals find themselves. Accordingly, goals can themselves be appraised or evaluated by the degree to which action on them does in fact tend to solve the underlying problem. Although such a formulation is cast in terms of the individual and employs bio-psychic con-

---

4. Especially as worked out in his *Human Nature and Conduct* (New York, Modern Library, 1930), *Theory of Valuation* (International Encyclopedia of Unified Science, II, 4, University of Chicago Press, 1939), and Part II of Dewey and Tufts, *Ethics* (revised edition, New York, Holt, 1932).

Attempts have been made in recent times to use a narrower concept which formulates ethical questions in terms of processes of decision and its factors. This orientation has proved useful in studies of administration, law and politics, and is being increasingly advocated in morals and ethical theory as well. See, for example, Herbert A. Simon, *Administrative Behavior* (New York, Macmillan, 1947); Wayne R. Leys, *Ethics for Policy Decisions* (New York, Prentice-Hall, 1952); M. R. Kadish, "Evidence and Decision," *The Journal of Philosophy*, XLVIII, 229-242 (April 12, 1951); R. M. Hare, *The Language of Morals* (Oxford, Clarendon Press, 1952). However, for considering the indeterminacy problem the wider mode of analysis in terms of problem-situations sets the scene more clearly.

cepts of impulse-habit configuration, it is readily generalized for groups and for extended time-spans in the historical life of a people. It thus provides an abstract framework in which the general indeterminacy problem may be considered.

We may thus speak of a problem-field constituted by the relations of persons, aims, circumstances, environment. In this generalized form there is no prejudgment as to the precise character of the constituents, no a priori limitation of the types of entities or events to be found relevant within it, nor of the respective importance of the different constituents in ethical action. We shall see that the contributions of the different sciences may then be regarded as explorations of the particular constituents and relations of the field from specialized perspectives.

In such a framework, the degree of determinateness in ethical judgment or decision depends to a large extent on how well-structured the problems themselves are in the total field in which they occur. The field is constituted by the relation of persons, aims, circumstances. But if the field lacks stability, if there is too great a flux in existence or too rapid a shift in people's aims or purposes, the structure will be so fleeting as to approach a structurelessness in the problem-situation. Similarly, too great a complexity in circumstances, too strong an inner conflict of purposes, may yield a plurality of structures within the situation which would make a determinate solution within the limits of the problem unavailable. Thus what we may call *field insta-* *bility* and *field complexity* are primary sources of indeterminacy. We shall have occasion to explore these sources at different levels—biological, psychological, social, historical —in our subsequent investigations, and to estimate their actual extent.

The question is even more complicated than it seems at first. For the indeterminacy may reflect not merely an *ante- cedent* instability or complexity, but one that develops in the process of decision or judgment itself. The very attempt to resolve a problem and the very consciousness involved in the process of solving it must be reckoned as possible participating factors and estimated for their stabilizing or disequilibrating influence, for their simplifying or complicating effect.

All these structural factors are "objective" in the sense that any indeterminacy to which they give rise need not reflect merely a lack of knowledge on our part. It is true that they have subjective consequences, but we may also look beyond these to the properties of the field itself. Thus great complexity means that the structure may be unmanageable to persons of limited powers; but it also indicates an objective multiplicity of elements. Similarly instability may yield confusion of purposes and frustration in selection of means, but it also refers explicitly to a high rate of change in the constitution or quality of the field components.

INADEQUACY OF CONCEPTUAL TOOLS AND GAPS IN KNOWLEDGE. Sources of possible indeterminacy are also to be found on the "subjective" side, in the traditional sense of shortcomings of the knowing process.

Concepts, like tools, may be too blunt or insufficiently refined. They may be so shaped as not to make close contact with the material, or if they do, lay hold on too much. For example, to say of a man at a given time that he is now having a "feeling of obligation," while it indicates something more or less recognizable, does not make close enough contact with the material to tell us exactly what he is feeling. On the other hand, to say that he has a "feeling of guilt" may be laying hold on too much and be inadequately refined for moral analysis. The whole question of the way a concept makes touch with existence has been widely explored by the philosophy of science in the theory of operationalism.[5]

Again, indeterminacy may stem from the lack of sufficient knowledge or information to permit predicting the outcome of alternative courses of action. In many cases this

---

5. Cf. P. W. Bridgman, *The Logic of Modern Physics* (New York, Macmillan, 1928), ch. 1. For a general summary of the philosophical problem, see the writer's *The Theory and Practice of Philosophy* (New York, Harcourt Brace, 1946), ch. 7.

further knowledge is taken to be objectively available although not yet "possessed."

This treatment of conceptual lack of clarity and unacquired knowledge as sources of indeterminacy holds within a certain range of human inquiry. But at its limits questions may arise as to whether such indeterminacy does not again reflect the primary structural factors—extent of change and complexity—which may put the clarification of concepts and the discovery of laws or description of existent conditions out of question. Thus concepts may remain insufficiently precise or unavoidably vague because of the primary richness and complexity of existence.[6] And the theoretical issue has often been raised in the human sciences whether the uniqueness of individual events may not involve, in some cases at least, an absence of the kind of analyzable regularities which alone make for "lawfulness" in nature.[7]

DIFFERENCE OF AIMS AND PURPOSES. The structural factors in the problem-field include the aims of the persons involved, and the conflict of aims is one of the chief bases of instability and complexity. As we saw in the chapter on the stubborn man, it is even assigned an ultimate status. Its place in the problem-field is clear; indeterminacy does often stem from conflicting aims. Even when the specific outcome of alternative courses is predictable different outcomes may be desired. Sometimes these differences arise even where there were identical aims in the original structuring of the problem; for conflicting interests may be evoked by the action that seeks to resolve the problem. But different aims, the lack of shared purposes, and even the presence of conflicting aims do not necessarily produce a high degree of ethical indeterminacy. Differing aims may require a common program of action to some extent,[8] just

6. Cf. Max Black on "Vagueness" in his *Language and Philosophy* (Ithaca, Cornell University Press, 1949).

7. Cf. M. R. Cohen, *Reason and Nature* (New York, Harcourt Brace, 1931), Book III, ch. 1; Franz Boas, *Race, Language and Culture* (New York, Macmillan, 1940), pp. 310-311.

8. Cf. C. L. Stevenson's analysis of patterns of agreement and disagree-

as a gang may observe a scrupulous code to get the booty before quarrelling over its disposition! Even Hobbes' state of nature, in his *Leviathan,* with every man's hand against every other, produced an ethic integrating differences and conflict into determinate ethical patterns of cooperation and competition.[9]

Nevertheless it probably is true that there would be a high degree of ethical indeterminacy if there were no bases for similarity of aims in men or for shared purposes in the conditions of their lives. To this extent, a great part of inquiry involves focussing attention upon such bases as the human sciences have explored them.

RELATION OF THE SOURCES OF INDETERMINACY. It is important to note that the phases distinguished as sources of indeterminacy—field properties of instability and complexity, inadequate concepts, insufficient knowledge, conflict of aims—are not wholly separate and cannot each be dealt with alone. Inadequacy of concepts may rest on lack of knowledge, some lack of knowledge may be ignorance of the very nature of our aims, while aims in action must themselves embody knowledge. And all of these reflect to some extent the primary structural properties of the field. At most we have here analytic phases of indeterminacy in ethical judgment.

§ Towards Coping with Indeterminacy

MODELS FOR COPING WITH INDETERMINACY. In approaching the problem of ethical indeterminacy it is useful to see what lessons can be provided by other fields that have faced comparable problems. In the wide range of available models we select two—one from the way in which indeterminacy has been handled in the history of physical science,

---

ment in his *Ethics and Language* (New Haven, Yale University Press, 1944), ch. 8.

9. For the question of reasonable rules to be followed, where everyone plays against everyone else, in order to ensure a maximum winning, see above p. 63 f.

the other from the way in which it has been narrowed down in legal theory. There is much to be gained in taking such apparently disparate models. One gives the initial appearance of a knowledge problem, the other of a practical judicial or volitional decision problem. The relation of such aspects in ethical theory has been a central issue in asking how far there can be determinate ethical knowledge to control ethical decision.

THE INDETERMINACY PROBLEM IN THE HISTORY OF SCIENCE. In looking at the magnificent structure of present-day physical science we are often prone to forget the range of attitudes that accompanied its development. Plato, for example—so Aristotle tells us—was convinced by Heraclitus that the material world is a constant flux. For Plato this meant quite literally that we could have no *knowledge* of the material world, for it was too indeterminate. He ended up with the view that it was almost wholly unreal. He did allow a kind of *opinion* or *conjecture,* since obviously we do move about with a little awareness of what is going on. A sympathetic modern critic will simply say that Plato was underscoring the now admitted probable character of empirical knowledge as against the certainty of demonstration in arithmetic, geometry, logic. But there is more to it, true as this is so far as it goes. For it represents an anti-empirical orientation for Plato. There is no real casting of the job of science in terms of creation of theoretical structures adequate to prediction and capable of advancing control. There is no central role for experimentation because nature is incapable of answering questions—she is herself too indeterminate. Hence Plato's stress falls almost wholly on magnificent theoretical structures released from responsibility to experience. These are the Ideas, eternal and immutable. These *norms* in mathematics, logic, morality, politics, are universals grasped by the intellect, while particulars are merely perceived. Plato argues for the mind's knowledge of them in a preexistent state, prior to birth, using his belief in the transmigration of souls to solve his epistemological problem.

What makes a present-day Platonic theory of science inadequate is not any inconsistency in his argument nor any narrowness in his scope. It is rather the actual *results* of empirical science which have pushed back the presumed indeterminacy of matter and brought large areas within the range of prediction and to a lesser degree control. As a consequence the Platonic orientation is reversed. Theoretical structures cease to be regarded as ends for contemplation or rulers exercising a divine right of kings over indeterminate matter. It is nature, rather, that in Santayana's phrase is free to bloom untrammeled. "Indeterminate" matter is free to exhibit her most eccentric whim, and theory like an elected official pledged to serve the subjects' welfare, must take some account of it. Hence eternal norms are replaced by alternative mathematical structures to be selected for their capacity for service in the understanding, prediction and control of actual phenomena.

Such a development resting on the results of science does not, of course, guarantee the complete determinacy claimed a priori in classical mechanism. In fact, it leaves open at crucial points a kind of choice. Causality implies that from the description of the initial state of a system (S_1) and the laws of the system one can predict accurately a subsequent state of the system (S_2). Now if S_2 is not successfully predicted, or in extreme cases if theoretical grounds are offered for an inability to predict it, as in Heisenberg's Uncertainty Principle in the theory of quantum mechanics,[10] two paths lie open. One is to place the indeterminacy in nature and deny that there are laws capable of being discovered. The other is to maintain the belief in the discoverability of laws and to place the indeterminacy in the inability to achieve

---

10. Max Planck, in his *Scientific Autobiography and Other Papers*, trans. by Frank Gaynor (New York, Philosophical Library, 1949), p. 133, formulates the essence of this as follows: "for any two canonically conjugate magnitudes, such as position and momentum or time and energy, only one can be measured to any desired degree of accuracy, so that an increase in the precision of the measurement of one magnitude is accompanied by a proportional decrease in the precision of measurement of the other. Consequently, when one magnitude is ascertained with absolute accuracy, the other one remains absolutely indefinite."

accurate descriptions of the initial state of the system. A complete elimination of indeterminacy would have to overcome not only temporary inabilities to achieve such accurate state descriptions, but also theoretically demonstrated inabilities. It would rest on the hope that whenever a branch of science acts as a basis for such a demonstration that branch will ultimately be revised. How realistic such a hope may be in any given area cannot, of course, be judged a priori.

From this brief picture of indeterminacy in physical science, important clues may be found for estimating the theory of ethical relativity. For example, classical moral absolutism seems to postulate an indeterminacy in the field of human nature almost like that postulated by Plato for matter. Plato, in fact, did make such a conception explicit in his theory of human passions; he regards appetite (in his *Republic*) as utterly arbitrary and capricious, capable of being restrained only by an external rational power. A well-developed theory of man provided by psychological and social science may yield the same reversal of outlook in regard to ethical norms that followed in the case of the interpretation of mathematical theory.

It is also important to realize, however, that even if systematic knowledge were developed in the various relevant areas, this would not automatically settle the issue how far determinate answers could be provided in particular cases. For example, algebra is a systematically organized field with laws, definite operations of solution, and so on. But this does not preclude two roots to a second degree equation; e.g., if $x^2 - 7x + 12 = 0$, then $x = 3$ or $x = 4$. Can ethics be satisfied with a conclusion that when two men are clinging to a log which can only support one, and their lives are of roughly comparable social value, either one should give place to the other? In such cases—that is, if in some areas it is found that no single answer can be reached—we should be left with the task of making major decisions of policy about what is to be done with the residual indeterminacy.

THE INDETERMINACY PROBLEM IN LEGAL THEORY.   On

the face of it, a legal system insists on complete determinacy. It cannot allow a decision that the disputed property belongs to either the plaintiff or the defendant. It will often try to plug gaps by procedural rules empowering the judge to decide in dubious cases, calling legal what the judge decides, and giving him directions about where to look for guidance; sometimes it even forbids a judge to refuse to decide on ground of inadequacy in the law. And in many ways it seems positively to adopt the indeterminacy by sanctioning whatever happens to happen in areas that it specifies. For example, procedures of mediation may be written into the law to have binding effect where used. Contract law really specifies that any promises of a certain kind made in certain ways will be assured or compensated for; this allows individual wishes to determine what the law will be in specific contexts in their own relations. Sometimes the parties are even left to fight it out with strength determining the result: for example, strikes are among the recognized modes of settling labor disputes today, and trials by combat occurred in olden days. In international law it is usual to recognize the winning side in a civil war, provided it sets up a stable government. Finally, where procedural methods do not suffice and indeterminacy may not be embraced, the law may escape worry by renouncing any pretension to universal scope. Some things may simply be beyond its jurisdiction; an extreme case is the concept of a "state of nature" in which no authority holds.[11] In the early history of English law, the limited role of law was clearer. Its types of cases were formulated under a limited set of writs, and if you could not fit your complaint in one of the pigeon-holes, you were simply not in the province of the law—literally, "no writ, no right."

Now such modes of limiting indeterminacy may seem to be verbal or arbitrary. But this is to overlook the real judg-

---

11. For an excellent hypothetical test case that sheds light on some aspects of these problems, see "The Case of the Speluncean Explorers," in Lon L. Fuller, *The Problems of Jurisprudence* (Brooklyn, The Foundation Press, 1949), ch. 1.

ment of value that may underlie the particular mode of treatment. Where the reaching of a decision is urgent— sometimes more important than what the decision will be —the procedural devices really serve to shut off dragged out conflict and dispute. The recognition of outside mediation procedures rests on the specific problem of overcrowded legal calendars as well as legal expenses. Contract as an institution expresses the major value judgment of modern business societies that a wide area in business relations be left to individual determination. Strikes, although costly modes of decision, are found by experience to be a necessary safety-valve in our society. Recognition of the winner in a civil war may avoid its extension into an international war. Even limiting the borders of law is not necessarily surrender or the confession of helplessness; it represents a value judgment concerning the role of law as an institution in a given age and place, and it implies that law may grow or contract as need for its unique contribution occurs.

Such modes of dealing with indeterminacy carry important lessons for ethical theory. Even the possibility that some decisions lie outside of the moral domain, important though they may be to the participants, is not precluded a priori; it involves, however, a fundamental policy decision on the nature and role of a moral pattern. At the very least, we see that indeterminacy of the particular does not preclude a large measure of determinate theory—that is, indeterminacy in a particular decision is quite compatible with determinacy in some of the factors of decision. For example, two men may agree that a close relative needs a home, but not on which of them should take him in. Even in the extreme case, the classic example of the two men in the water clinging to the log, a number of moral agreements are possible. Certainly one man ought to relax his hold, since to save one life is better than to lose two. (We leave aside special cases of two friends who might decide to die together, or two excellent swimmers who might indefinitely take turns, using the log as a resting-place.) There might be agreement on more specific values—let the one who has a family survive, or who has a more important social role; or let the one who has lived a fuller life be ready to leave it. If nothing else solves it, there may be common methodological values—let it be settled by some chance method, or by strength, or by each trying to sacrifice himself. However irrelevant such considerations may seem to the participants, they may very well be relevant principles for education or advance training, for after-praise and after-blame, or for social welfare. Such judgments may be comparable to probability judgments on a frequency theory; that is, they refer to the outcome in classes of cases rather than to individual cases. But their own scientific character is not diminished thereby.

How Contrasting Are the Models? In the physical science model, stress fell on the growth of factual knowledge about existence, making possible greater prediction of what will happen in particular situations. In the legal model, stress fell on the role of shared purposes in limiting indeterminacy in the system in giving answers concerning what is to be done. There is a strong contemporary tendency in ethical theory to draw a sharp contrast between the descriptive and the prescriptive, the scientific and the imperative or normative. Thus it will be said that the scientific model shows how we *find answers* to questions of fact, whereas the legal model show how we *make decisions*. In the former we are learning, in the latter we are in some sense doing or creating.

In many ways, as we saw,[12] this sharp fact-value separation has become one of the pillars of ethical relativity, since it is assumed that questions of fact admit of scientific answers in principle, whereas questions of value rest ultimately on some arbitrary act of throwing in one's lot. Such issues have deeper roots than is often recognized. This familiar distinction assumes at bottom a sharp separation of intellect and will. Therefore it should not be accepted at the outset

---

12. See above, pp. 74 ff.

as if it were clear on the face of it that scientific decision is
an act of knowing and legal decision an act of willing. There
are creative elements in science and there are cognitive elements in law. It is possible to distinguish between the *act*
of deciding and the *cognitive content* of the decision. We
can construe the legal judgment as a cognitive conclusion
that such-and-such is according to or contrary to a particular legal system, and see the fiat element, "Be it so ordered,"
as a volitional supplement. Now whether the scientific and
legal model should be drawn together or sharply distinguished is thus not settled by initial possibilities of constru-
ing one way or another—for both are possible—but by careful
examination of how the models function and by evaluating
the consequences of both courses.[13] From this point of view
it is at least worth entertaining the hypothesis that the difference between the models is one of degree.

Such a hypothesis means that both models involve comparable phases of the decision process. Differences between
them are fundamentally differences in emphasis representing such factors as: degree of standardization of problems,
concepts and shared purposes; degree of determinacy of
results; differences in actual materials involved, such as
greater complexity of human qualities as compared to physical qualities. Thus both will have governing purposes and
guiding ideals in inquiry, although those of physical science
are by now almost wholly standardized (prediction, control,
conception of system), whereas those of law (security, justice) remain controversial. Both employ abstract conceptions, although force, energy, and field conceptions will be
precisely analyzed whereas conceptions of due process, tort,
even rights, will be vague and fuzzy. Both employ generalizations and principles, although those of physical science
are immeasurably more established and certified. Both invoke criteria of relevance for applying principles to particu-

---

13. I have discussed this question in the fuller context of contemporary trends in ethical theory in "Ethical Reasoning," in *Academic Freedom,
Logic and Religion*, ed. M. White (Philadelphia, University of Pennsylvania Press, 1953).

lar situations, although there is considerably greater leeway in regarding a phenomenon to be explicated as one of contract or property than as one of mechanics or electricity. Both employ operational tests, although the verdict of twelve men good and true, or the overruling of a lower court by a higher court utterly lacks the refinement of pointer-readings in designed instruments. Both have conceptions of desirable type of system to be achieved, although one is developed in centuries of refined analysis while the other muddles along with rough models.

And yet if we looked less at physical science and more at psychological and social science, the differences might be narrowed. The controversies are familiar enough. Can the social sciences properly guide their inquiry to achieve prediction, or should they aim at some interpretive appreciation? Have they clear abstract conceptions? (What, for example, is a culture or a social system, or a value, for that matter?) Are there social laws? Can the relevance of a set of generalizations to a particular phenomenon be readily determined? (What is cultural, what psychological, what physiological, in a particular happening?) Are there clear operative indices for the existence of, say, a democracy? And so on. Yet one would not question that the social sciences are "factual" sciences. What this comparison shows then is that many of the distinctions between the science model and the legal model have nothing to do with the fact that one is "descriptive" and the other "normative" or "prescriptive." They concern rather degree of precision in concepts, extent of accumulated knowledge warranting generalization, and degree of established guiding aims.

Yet it is true that in law the normative or prescriptive element looms large whereas in science "finding answers" does not seem to have the same volitional character. The scientific proposition does not seem to demand that you believe, as the judge's seems to demand that you do. I am not suggesting that we overlook any actually discoverable differences between the two models, but that we do not make an absolute distinction where the difference may be one of

degree of emphasis. And the possibility seriously worth considering is that in decisions which consist in "finding answers" in the descriptive sciences the prescriptive elements
are either already built into the core of the scientific enterprise so that we take them for granted (such as the pre-
dictive aim), or else pushed to a peripheral position (as in
the control motif), or both. In the normative disciplines
they have a relatively more central place because they find
a home in the greater fuzziness of concepts, in the lack of
authoritative factual assumptions, in the conflict of proposed
principles, and so on. Thus the prescriptive element seems
to enter more detailedly into the particular judgments rather
than be set in the background framework in which the judgments have a systematic place.[14]

The study of indeterminacy in ethical judgment can thus
learn from both models. It can see what is required in the
clarification of concepts, in the extension of underlying
knowledge of man, in the standardization of basic purposes
in its enterprise, to give it a position closer to the present-
day science model. Similarly it can see what conditions
would put it closer to the present-day legal model. But
where it actually lands and the degree of indeterminacy that
prevails cannot be resolved by mere wish or fiat, nor argued
by analogy from either of the models. It requires an appeal
to the actual concepts that ethical theory has cumulatively
developed, the actual knowledge that the human sciences
have by this time established and the actual basic shared
purposes that men possess by endowment or cultivation.

SUMMARY AND GENERAL HYPOTHESIS. Our inquiry into

---

14. A comprehensive investigation is required to map the exact points
in a descriptive science or normative discipline at which the prescriptive
element may be located, how it may shift from point to point and under
what conditions, and what degree of freedom there may be, while sys-
tematizing a field, in locating it one place or the other. This would raise
questions of the desirability of locating it in one place or the other, and
why. In many respects such an investigation would be parallel to the
study of the location of indeterminacy indicated above in the pecking
order of theories. They are not unrelated, insofar as the points of prescrip-
tion are also the points of choice of standards and so also the points where
indeterminacy may be found to cling.

what ethical relativity is as a theory began with the recurrence of the sense of ultimate arbitrariness in the formulations of contemporary problems. We mapped the several threads entering into the design that supports this feeling. We found the unity of the design to lie not in the fact of variety and potential conflict but in a stress on basic indeterminacy in ethical judgment. That is, as a doctrine, ethical relativity makes indeterminacy the center of gravity in ethical theory. As so formulated, the problem of ethical relativity can no longer be dealt with by an off-with-its head policy, nor again by simply assuming that it is the last word of science. It requires formulation as a *research* issue, not as a subject of postulation one way or the other.

A brief examination of the way in which various ethical theories handled indeterminacy suggested that it found a home in many different elements of a theory, and that when it was routed out of one place it took refuge in another. The attempt to rule it out or to enshrine it by assumptions of fact or stipulation of method were found to leave gaps which only scientific knowledge can hope to bridge.

The general sources of indeterminacy were then investigated, and contrasting models for its treatment considered. Here although such aspects as inadequacy of concepts, lack of knowledge and conflict of aims could be distinguished, the basis of indeterminacy was found to lie in the way in which they were related in the actual or objective conditions of the problem field. Hence the fuller study of indeterminacy carries us to the contributions of the human sciences in understanding this field and estimating its degrees of instability and complexity.

The complete elimination of indeterminacy is, of course, by no means the aim of our present inquiry. To pursue this in human affairs is an idle dream. But it is quite possible that some of the strands in the design of ethical relativity which were most influential in engendering indeterminacy have given way in the growing knowledge of man. If so, the central emphasis on indeterminacy may well represent an early and hasty generalization. That their replacement by growing contemporary knowledge reverses the direction in ethics is a guiding hypothesis in this book. We shall find that an ethics utilizing science is making great strides. Shortcomings have arisen from concentrating on one or another of the various scientific perspectives, but the results are already adequate to remove indeterminacy from the position of center of gravity in the ethical sphere. Although there is no promise of its complete elimination, enough has been accomplished in these sciences to get current ethical bewilderment in hand, and to deal with the "uncertain" factor by a participant and creative approach in social and individual life. Major decisions of policy become possible which lose the character of arbitrariness, and residual indeterminacy is hemmed in by well-grounded values so as to be no longer a general problem but a set of particular questions requiring separate decision. If such major policy decisions are possible, supported by an interlocking structure of human knowledge and human striving, then we may well think of them as *moorings* to which morality is fastened and from which it cannot drift very far, in spite of variant motions within definite bounds. There may still be sharp tugs as winds rise, tossings up and down, plenty of seasickness; but the moorings are firm and we know where we are.

How far such a program in ethics can really be carried out depends then upon what we can learn from the various sciences about the structure of the problem-field in which human ethical judgment takes place. To explore these contributions of the sciences from biology to history and carry out a philosophical analysis of their impact on ethics is the task of Part II of the book.

PART TWO

What the Human Sciences Can Offer

V. Biological
Perspectives

The perspectives we are to examine may be organized according to the roster of the human sciences, beginning with biology. To the purely physical sciences we need not go, although much might be learned there concerning methods and guiding ideas. Nor again need the distinction of biological, psychological, social and cultural, and historical be taken to correspond to any sharp divisions in the actual life of man. They constitute rather accumulated bodies of knowledge and modes of inquiry organized around certain foci of investigation. Their relations and the borderline areas between them are also important, and in the long run attention to them may yield greater unification of perspective. But at present we are looking for the impact of accumulated knowledge on the problem of ethical indeterminacy, and so our inquiry is shaped to the contours of that knowledge as it has historically developed and at present exists.

§ Attempts to Derive Definitive
Moral Patterns from Biology

MAJOR DIRECTIONS. It is fairly evident by now that biology has been overworked, especially during the past century, in the attempts to provide a "scientific ethics." In fact some of the disrepute into which this concept has fallen in many quarters stems from the haste with which whole moralities were erected on general biological theories.

The most prominent and spectacular of these moralities were evolutionary, rising to a crest with the popularity of Darwinism. There seems little or no limit to the variety of moral lessons that men have read on the basis of the sober facts of biological evolution. In addition, many have looked to biology for an account of the "ultimate" nature of man, seeking a set of fixed characteristics on which to build a determinate morality. Biology has also provided perennial models, such as the organism concept, on which ethical theories have been patterned.

EVOLUTIONARY MORAL ORDERS. The mainstream of 19th and early 20th century ethical interpretations of evolution sang the praises of struggle as the instrument of upward movement. Herbert Spencer's *Social Statics* which (as Spencer pointed out later) actually antedated Darwin's *The Origin of Species,* launched an era in which the *struggle for existence* and the *survival of the fittest* were to become major slogans of a predatory business ethics. The sufferings of the poor and the miseries of the oppressed were seen as by-products of a beneficent evolutionary process in which the able came to the top and the unfit were wiped out. Any interference with this self-propelling upward and onward system of struggle and competition was evil, an absolute violation of nature's moral law. The influence of this whole constellation of ethical justification in American life may be judged from Holmes's dissenting opinion in the Lochner case, in which he opposed this outlook, stating bluntly that the 14th Amendment was not intended to enact Herbert Spencer's *Social Statics* into American law.[1] The law which the Supreme Court struck down as unconstitutional had

---

1. *Lochner* v. *New York* 198 U.S. 45 (1905). For a study of the manifold impact of the evolutionary modes of thought in American social theory, see Richard Hofstadter, *Social Darwinism in American Thought 1860-1915* (Philadelphia, University of Pennsylvania Press, 1945).

simply attempted to limit the working hours of bakers in New York as a public health measure.

Competing patterns of interpretation for nature's moral order have taken many different paths. Kropotkin marshalled impressive data for the role of cooperation in survival and development, and tried to see the whole of human history in terms of a rising tide of mutual aid.[2] Many, seeing the growth of larger and larger cohesive human social groups, have read evolution as justifying a coming universalism. Others see the perennial conflict of in-group and out-group and draw moral lessons of the need for group independence through strength and readiness for war.[3] Others search for persistent evolutionary trends such as increased range and variety of adjustments of organism to environment,[4] and are ready to equate the more evolved with the humanly better.

SEARCH FOR "ULTIMATE" INSTINCTS.   Those who look to biological accounts of the organism and its make-up for a moral ground-plan, often come to rest on some fact of fundamental drives or instincts as the ultimate nature of man. The underlying assumption is that an ethics has to take man for what he is. To be rooted in a biological mechanism serving a biological need seems to many to be all the justification human behavior can ultimately ask. In this way, eating and drinking and sexual activity are obviously sanctioned. A variety of other invariant human tendencies, it is believed, will be found similarly to rest on more subtle biological mechanisms and thus prove their merit. Any institution claiming roots in instinct can thus clothe itself with the moral authority of absolute fixity. The history of social psychology is strewn with the wreckage of instincts intended

---

2. Peter Kropotkin, *Mutual Aid* (Penguin Books). Darwin had himself, in *The Descent of Man,* underscored the values of sympathetic feelings in human society as major survival aids; he even regarded these bases of the moral sense as instinctive.

3. E.g., Sir Arthur Keith, *Evolution and Ethics* (New York, Putnam's, 1946), pp. 145-147, 192-193.

4. For a brief study of such interpretations, see, G. Gaylord Simpson, *The Meaning of Evolution* (Mentor Book, 1951), chs. 7, 10.

to support prevalent institutional forms, such as pugnacity instincts to prop up war, acquisitive instincts to reinforce private property, and a variety of specific instincts to support the family.[5]

USE OF BIOLOGICAL MODELS. Some attempts to give determinate shape to ethical theory employ organic processes as *models*. Most common has been the treatment of society as a kind of organism, followed by elaboration of a concept of social health and a set of moral precepts required for maintaining this health and well-being. Leslie Stephen, for example, in his *The Science of Ethics* (1882) uses the language of social tissue and its vitality. The most influential recent model is Cannon's concept of *homeostasis*.[6] Primarily, it refers to the maintenance of a steady state—for example, in temperature or in sugar concentration of the blood. Regulatory processes come into action to preserve the state as it reaches one or another extreme. There has been some tendency to apply this concept in social theory; thus the maintenance of a social order or a cultural pattern can be seen as a kind of social homeostasis. Rights and duties can be defined in such a conception by seeing the regulations functioning to maintain the steady state.

GENERAL EVALUATION. These various attempts to derive absolute or determinate moral patterns from biology in the form of evolutionary orders, instinctual constitutions or process models, raise serious problems of evaluation. They may be estimated briefly along three main lines—the clarity of their concepts, the factual truth of their assertions, and the desirability of their implicit valuations.

Many of the concepts employed in the several accounts are not as clear as they seem on the surface. Darwin himself had some qualms about the meaning of the term "struggle for existence." It clearly assumes a quite different character where it refers to a plant struggling against drought, men

---

5. Compare the way in which instincts are treated in William Mc-Dougall, *An Introduction to Social Psychology* (Boston, J. W. Luce, 1908), and Otto Klineberg, *Social Psychology* (New York, Holt, 1940).

6. W. B. Cannon, *The Wisdom of the Body* (New York, Norton, 1932).

struggling against disease, individuals competing with one another in a socially structured set of activities, different human groups making war against one another, or one species using another for food. Similar issues arise in the meaning of "cooperation," "adaptation," etc. And further complications of meaning arise when it is asserted that struggle or cooperation have been "primary" in the evolutionary process. Nor have the criteria of what rates as "instinctive" nor what constitutes an "organism" or "organic whole" been elaborated with the precision that would warrant their ethical application.

Examination is also required of factual assumptions. There have been numerous controversies about the descriptive picture of evolutionary processes, mechanisms of transmission, conditions of group or type stability and change, and so on, whose answers would have a quite varying ethical potential. Moreover, it will be found that on the whole ethical conclusions in the apparently biological ethics use anthropological and historical as well as biological premises,[7] and these too would have to be scrutinized for their accuracy.

Beyond the conceptual and factual issues lie the problems of implicit valuation in the several forms of biological ethics. To what extent can the basic activities focussed upon in the various conceptions—life, struggle, cooperation, individual or group expression, themselves be regarded as good rather than as neutral or in some cases evil? Many volumes of criticisms in recent decades have made clear the distinction between finding an evolutionary order and going on to declare it a moral order or a progressive order—obviously the world may be going downhill. These criticisms have, it is true, sometimes overlooked the complexity of the theories which at their best were not just tracing an order of events but describing the evolution of valuations themselves within

---

7. This is explicit in Spencer. Recent discussions such as Julian Huxley's in *Touchstone for Ethics* (New York, Harper, 1947), or Ashley Montagu's in *On Being Human* (New York, Schuman, 1950) make considerable use of contemporary psychological materials.

the order. Nonetheless, especially in the light of all this criticism, an evolutionary ethics does require an explicit account of its value components. Similarly, the call for instinctual expression, maintenance of a steady state or adaptation to a given set of conditions must show the grounds for assuming the goodness of the activity, state, or results. Why not call instead for repression, change of state, change of conditions?

Now the mere multiplication of issues is not itself fatal to the hope of a self-sufficient biological basis for a determinate morality. The vague concepts might be clarified, the factual issues might be further explored and adjudicated, the valuational standpoint stabilized. And the past excesses might be seen as ideological use of biological data constituting distorted interpretation. But these possibilities do not seem enough. In fact the more biological materials are developed, the clearer it becomes that ethical judgment is dependent on the coupling of biological data with that furnished by the other human sciences. We shall see even in this chapter that the simplest values propounded on biological grounds require justification in part in at least psychological terms. This suggests that we should no longer look to the biological perspective for an independent ethics but rather for foundations and at most a partial framework.

FUNDAMENTAL CONTRIBUTIONS TO THE PROBLEM OF ETHICAL RELATIVITY. Once the search is narrowed down in this way, we find that a biological perspective has made and can make a vast contribution to our subject. For if man, in the old Aristotelian definition, is a rational animal, biology has given the picture of his animality on a wider canvas than any other branch of human knowledge. Briefly then, I should like in this chapter to consider the following contributions of the biological perspective to analyzing ethical relativity. It exhibits man as part of nature, made of the same common clay as other living beings, and points to the animal basis underlying the human spirit. It thereby imposes on us the task of carrying out a fuller "naturalization" of moral phenomena themselves. Second, it lays foundations, together

with psychology, for a scientific evaluation of life, pleasure
and pain, as basic referents in ethical judgment of variable
forms of human activity. Third, it brings into the center of
ethical attention man's biological impulses and drives and,
by applying to these its evolutionary theory, furnishes a mode
of evaluating their role. Fourth, as a consequence, it makes
possible the adoption of health as an evaluative criterion, and
provides a groundwork for a general evaluation of frustra-
trion. In addition, it lays the foundation for two important
analytical tools in the whole range of the human sciences
which are central to ethical evaluation. In the first place, it
develops and sustains the insight that moral phenomena always served some *functional* role, an insight which has been
confirmed and extended by the psychological, social and historical sciences. Secondly, by its very long-time perspective,
looking at the struggle of populations for survival, it em-
phasizes a group orientation in formulating ethical issues.
This proves a helpful corrective to some of the individualist
formulations that have dominated the modern scene.

§ The "Naturalization" of Moral Phenomena

RESTORATION OF MAN TO NATURE.   Biology has restored
man to nature. It has shown man to be definitely part of the
animal kingdom, evolved over a long period of time from
simpler forms. This revolutionizes a previously dominant
supernaturalist ethics. It means that man's animality is not
to be placed outside of the human who is the central figure
on the ethical stage, nor is it merely an obstacle to be tran-
scended. What precisely it consists in, what drives it entails,
what qualities it brings to human experience, what conflicts
it gives rise to, are to be discovered and integrated within
the framework of ethics.

On the other hand, such a recognition does not mean the
simple *reduction* of man to animality. It does not justify the
transfer to human ethics of the types of order found in jun-

gles or ant colonies, or even carrying over to present ethics orders that may have been invariant under early conditions of human life. A fuller evolutionary outlook does call for the naturalization of man's spiritual life but only in the sense of showing at what points and under what conditions the vast qualitative proliferation of specifically human phenomena occurred.

RISE OF MORAL AND ETHICAL PHENOMENA. In ethics, the outline is becoming clearer today. Drives, demands, needs, selections—in short the striving to live and grow and what in the widest sense we call *valuing*—in some form ante-date the appearance of man on earth. Of these, fundamental drives and needs, to some extent at least, exist as a consequence of biological constitution, and therefore require explanation in terms of the early evolutionary development of the human species. On the other hand, ideals and aspirations are obviously more specialized human phenomena built on these bases in a cultural milieu.

The sense of *obligation* in men seems to be a psychological and social product renewed in each generation out of human relationships, and not biologically transmitted. Some of the raw materials for this process—the binding phenomena of sympathy and the feeling of remorse, for example, seem to be found in rudimentary form in the higher animals.[8] But the finished product is clearly a human accomplishment, whether looked at in terms of the specialized qualities of feeling, or in terms of interpersonal relationships.

Evaluating has, of course, always been implicit in deliberate preferential judgment or in configurations of individual and group life-goals, just as length measurement and counting are implicit in judging one thing to be greater than another or one group to be larger than another. As a specific explicit process it appears to have flowered with the growing material means and increased stability of the species on the globe. This gave greater significance to choice and prefer-

---

8. For an illustration of some of the difficulties in these judgments, see S. Zuckerman, *The Social Life of Monkeys and Apes* (London, Kegan Paul, 1932), ch. XVIII.

ence and involved more conscious development of standards. It brought greater variety and specialization into social life, in which evaluating took shape as working standards of accomplishment and human relations.

In its various theoretical forms, *ethical inquiry* is a much later product, clearly tied to the historical pressures of civilization which intensified consciousness in its religious, artistic and philosophical expressions.

All these—from brute demand to most refined analysis of ethical language—are qualitative processes now going on in the life of man. Their study involves all the sciences of man, not merely biology, and the forces underlying them—whether described in the language of biological impulse seeking outlet, of psychological tensions and needs, of cultural and social pressures, or of historical groups and interests, or in some generic term on all levels such as "problems" or "conflicts"—should not be assimilated to the models of any one specialized discipline.

RETICENCE IN CARRYING OUT THESE TASKS. The details of this historical picture are still sketchy, not wholly because of lack of data, but more because of a prevailing unreadiness to naturalize ethics entirely. This task would seem to be enjoined upon those who accept wholeheartedly the Darwinian revolution. But it is halted half-way by entrenched and residual attitudes of the pre-Darwinian world. A glance at scientific journals is sufficient to show that biologists are often still fighting for the right to naturalize ethics rather than carrying out its naturalization in an evolutionary spirit. And academic ethics is sometimes even further back. Its ethical stage is still full of transcendental ghosts and pure minds intuiting ethical essences. Or else pleasures and pains attract and repel each other as little atoms of self-justified being. The wider evolutionary or historical context is peremptorily dismissed as purely causal, and therefore ethically irrelevant, or at most as exhibiting the relativistic variety of moral phenomena.

Such an attitude ignores the extent to which the picture of man's constitution and development contributes to the establishment of value foundations in an ethical theory. This can be seen in the evaluations of life, pleasure and pain to which we next proceed.

§ The Value of Life

EVEN OBVIOUS VALUES MAY BE QUESTIONED. The value judgments we are now to consider are of the type that most men will readily accept. Few will deny that pain and break-down are evil, that life, health and pleasure are good. Yet these judgments have been questioned. There are the extreme forms of eastern philosophy which disparage completely all worldly existence, and set as the good a Nirvana in which the spirit even loses all sense of individual being. Man is trying to escape from the wheel of life, and to value health or pleasure is to tie one's spirit to the earth. But even in more familiar terms, one may be inclined to say that it all depends: some pleasures yield evil consequences and some pains good consequences. Moreover, there are cases where pains seem sometimes to be sought, even glorified and institutionalized. We cannot just say of these with Herbert Spencer—after referring to blood-thirsty and tor-turous customs, and recognizing that devil-worshippers are not yet extinct—"Omitting people of this class, if there are any, as beyond or beneath contempt. . . ."[9]

Is LIFE A GOOD? Since the biological perspective thinks in terms of survival, let us begin with *life*. Is life a good? Or is it indifferent or even evil? Suppose that an actively-minded Schopenhauerian proposed to an already achieved world federal assembly, in appropriate constitutional form, that "this assembly recognize and declare that life is an evil, and take due steps to bring about, by use of the mani-fold methods—atomic and biological—now available, the termination of life upon this globe."

What rebuttal shall our global statesmen of the forward-looking type, whether from the capitalist or the communist

---

9. *Principles of Ethics* (New York, D. Appleton and Co.), Vol. I (1896), § 10.

sectors of the world, today offer to this resolution? Shall they call on the sergeant-at-arms to strait-jacket the honorable member as preparatory to committing him for lunacy? Shall they rule it out as universalized genocide? What about a procedural argument that this requires a global referendum? Or better still, it is undemocratic, since even if a majority wishes to end life, it cannot force a minority to do so. Therefore, at most, the honorable member is proposing to legitimize suicide. This becomes a more familiar theme, and it is easy to point out that while simple suicide has been condoned by Stoic moralists, no honorable suicide conducts his departure in such a fashion as to kill the unwilling.

Even the proposition that life is a good has sometimes met with the answer "It depends . . ." It depends on the amount of life and overcrowding. It depends on the quality of life. It depends on the attitude of the living beings; surely some hate their lives and even end them. On the other hand, an Albert Schweitzer goes from the Will to Live directly to an ethical principle of Respect for Life. And Tolstoian sects, such as the Doukhobors, practice vegetarianism and some even have refused to use insecticide on potato bugs threatening the crop, preferring instead to gather the bugs with infinite patience and carry them away from the field.

EVIDENCE FOR AN AFFIRMATIVE ANSWER. If we remember the broad definition of "good" in terms of striving and aspiration which was accepted earlier as a preliminary formulation, it becomes possible at least to recognize the scientific character of the question whether life is a good. Here the evidence provided by the biological perspective lends a basic bias towards an affirmative answer. All forms of life are seen as struggling to maintain and extend themselves and to continue life. The intensity and persistence of the effort toward survival on the widest scale is such that one is not surprised to find philosophies making respect for life the basic ethical commandment. On the contrary, one is surprised to find the occasional Schopenhauer recommending extinction, and one begins to look for special circumstances to explain his pessimistic reaction. But perhaps the reverential philosophies also constitute a reaction to men's widespread disrespect for their own and others' lives and powers; hence perhaps this sanctification of life as such is dictated by historical and psychological rather than simply biological forces.

INSTRUMENTAL KILLING OF OTHERS NOT A DISPROOF. In the face of the tremendous drive for life, the evidence that shows a negative or indifferent valuation requires careful critical consideration. It is possible that apparent exceptions to the value of life constitute not a *denial* of its values, but rather an *over-balancing* of its value by other aims to which termination or depreciation of life may be instrumental. Thus the vast panorama of evolutionary struggle and killing is not in itself evidence that life is an evil. The very intensity of the struggle and the fact that species uses species as food indicate that it is literally a struggle for survival, that is, for life. The same is true of killing in human affairs. Although as civilization has advanced technically, it has become less killing for survival than for resources, for power, or higher levels of material existence, in general, killing has had an instrumental character. There are very few cases that appear to be ruthless killings just for the sake of killing.

THE PHENOMENON OF SUICIDE. The same analysis, up to the same limiting point, applies to suicide as a negative valuation of life. Sometimes suicide is a realistic instrument: a man may end his life as a means to achieving some other goal. No assertion that life is good, as a phase-rule, assumes that it is the sole good or even the highest good. Thus a man captured by his enemies may kill himself because he knows he will not be able to resist torture without betraying his confederates. A man with a lingering painful hopeless disease may kill himself to avoid the pain and the burden upon his family. On the more positive side, a man may embark on a course of action which he realizes will bring his death, in order to advance at some critical point, a cause with which he has identified himself. In all these cases, the goodness of one's own life is weighed, but outweighed.

Usually, however, suicide is not so overtly realistic. It occurs in many cases of depression, sometimes accompanying well-defined psychotic or psychoneurotic states; and some psychiatrists accept the occurrence of impulsive suicide. For example, Henderson believes that "the aggressive act of suicide is, in many instances, a more individualistic response which may be elicited, almost instantaneously, in face of circumstances which may not have been easy to cope with by a personality who, throughout life, has reacted in a turbulent, impulsive manner to whatsoever difficulties existed."[10] Such impulsive suicide has the instrumental character of a rebellious attempt "to hit back." The predominant depressive type may, however, exhibit men actually hating their own lives. Their self-destruction has the character of aggression against themselves. To decide whether this is instrumental requires a fuller analysis of the phenomenon.

Contemporary Freudian psychology has revealed the full scope of such phenomena, of which suicide is only the extreme case. For example, Karl Menninger, in his *Man Against Himself*, provides a formidable array of evidence that actual men do in varying degrees punish, hate, maim, as well as destroy themselves, that this often has a bitter and vindictive quality, and that men are not always conscious of what they are doing to themselves in this respect.[11]

Both killing for what appears to be killing's sake and self-aggression thus constitute an indeterminacy point in the argument that life is a good. If, in spite of appearance, these are instrumental to other aims, they do not constitute a disparagement of life as such. If they are goals on their own, the goodness of life is dependent on the person's aims and life is good not universally, but only for those (majority or minority) whose will is directed on life.

THE THESIS OF THE DEATH INSTINCT. The chief attempt to give self-aggression and aggressive phenomena generally

---

10. D. K. Henderson, *Psychopathic States* (New York, Norton, 1939), p. 47. Cf. his whole discussion of suicide, pp. 46-56.
11. Karl A. Menninger, *Man Against Himself* (New York, Harcourt, Brace, 1938).

an independent status in human life fully coordinate with the will to life is found in the theory of a death instinct. The thesis that there is a death instinct (*thanatos*) alongside of a life instinct (*eros*) was advanced by Freud in order to unify the phenomena of hostility and aggression. When he first introduced it in his *Beyond the Pleasure Principle,* his eye was on a few special phenomena, such as the degree to which painful experiences are repeated as experience rather than as memory in the psychoanalytic process. Once he had assumed this "repetition-compulsion," he interpreted it as an innate tendency in living organic matter to return to an earlier state from which external forces had disturbed it. And he even toyed with the idea that pleasure, because it arises in the discharge of tensions, is really helping the death instinct keep down complication of the living matter.[12]

In a later book, Freud propounds the view, resting on the death instinct, that "we have to destroy other things and other people, in order not to destroy ourselves, in order to protect ourselves from the tendency to self-destruction. A sad disclosure, it will be agreed, for the Moralist."[13] In such works as *Civilization and its Discontents* and *The Future of an Illusion* we find a strange mixture of hope for the brotherhood of man and the reduction of suffering, expectation that in the long run nothing can withstand reason and experience, and at the same time a pessimism apparently resting on the inevitability of hostility and aggressiveness in man. Thus at the end of the *New Introductory Lectures on Psychoanalysis* he concludes after looking at Marxian theory and the revolution in Russia that even if new discoveries increase our control over nature and make easier the satisfaction of our needs, "we shall still have to struggle for an indefinite length of time with the difficulties which the intractable nature of man puts in the way of every kind of social community."[14]

---

12. Sigmund Freud, *Beyond the Pleasure Principle,* trans. by C. J. M. Hubback (New York, Boni and Liveright), pp. 24, 44, 83.

13. *New Introductory Lectures on Psychoanalysis,* trans. by W. J. H. Sprott (New York, Norton, 1933), pp. 144-145.

14. *Ibid.,* p. 248.

It is not surprising, therefore, that there has been a strong tendency in recent years to use the Freudian analysis of aggression as an ideological weapon, to substitute an invocation of the death instinct or the inevitability of aggressiveness for the concrete analysis of social causes in social phenomena such as war, to relegate attempts at social reform or revolutionary reconstruction to psychological Utopianism, and to condemn radicalism wholesale as aggression against the father. In effect, in spite of Freud's insistence that psychoanalysis is not a *Weltanschauung*, a world outlook, but only a specialized branch of psychology,[15] it has almost been made today into a type of apologetic ideology for a chaotic society.

On the whole, however, as a scientific theory, the death instinct has not won general acceptance since Freud advanced the view, even among Freudian analysts. The mere fact that dying is, so to speak, a constant part of biological living does not ensure it a particular psychological role. Again, too many of the phenomena of aggression—certainly those among nations or social classes or groups—are clearly social and historical phenomena not capable of explanation in terms of individual psychology. Explanation even of the individual phenomena is possible in other terms. Thus Fenichel, who is generally regarded as an authoritarian spokesman of the contemporary scientific Freudian school, says: "Of course, the existence and importance of aggressive drives cannot be denied. However, there is no proof that they always and necessarily came into being by a turning outward of more primary self-destructive drives. It seems rather as if aggressiveness were originally no instinctual aim of its own, characterizing one category of instincts in contradistinction to others, but rather a mode in which instinctual aims sometimes are striven for, in response to frustrations or even spontaneously.

"Aims are sought more readily in a destructive way the more primitive the maturation level of the organism—per-

---

15. *Ibid.*, ch. 7.

haps in connection with the insufficiently developed tolerance toward tensions."[16]

Again, in the light of the social usage of the Freudian
account of aggression, it is worth noting Fenichel's conclusion on prophylaxis: "Neuroses are the outcome of unfavorable and socially determined educational measures, corresponding to a given and historically developed social milieu
and necessary in this milieu. They cannot be changed without corresponding change in the milieu.

"If a society becomes unstable, full of contradictory tendencies, and the scene of struggles between its different
parts, power alone determines how and toward what goals
education is directed. The instability and contradictions of
a society are reflected in its education, and later in the
neuroses of the educated individuals."[17]

If the denial of the death instinct is correct, and if both
killing or hurting others and self-aggression represent more
or less distorted ways of striving for other aims, rather than
themselves being ends in human consciousness, they cannot
constitute evidence against the judgment that life is a good.
They may be means the individual uses to diminish the
burden of anxiety or guilt-feeling which has fastened its
hold over his whole mental climate, inflicting one evil to
remove another evil. But even if the guilt-laden atmosphere
turned out to be a pervasive feature of all human life, so as
to make it on the whole miserable, it would not for that
reason alone disprove that life as such is a good. It would
be the same as if some pervasive condition of the physical
atmosphere emerged which made the human body suffer
continual irritations and burns up to the point of making
life on the globe intolerable. One might then conclude that
although as a phase-rule life is a good, the actual results of
the reckoning under such universal global conditions would
be invariably negative.

ATTITUDES TO DEATH. Not too much can be determined

---

16. Otto Fenichel, *The Psychoanalytic Theory of Neurosis* (New York,
Norton, 1945), p. 59.
17. *Ibid.*, p. 586.

at this point for the value of life from men's general attitudes
to death. In any case, death is a quite different problem
ethically, and its evaluation is by no means a direct function
of the value assigned to life. There is no inconsistency in
holding that life is a good and death is indifferent if one's
life has satisfied certain qualitative conditions. From the
wider biological perspective, as Corliss Lamont points out
in *The Illusion of Immortality,* death serves a salutory evolutionary role in clearing the stage and permitting the development of new ways and new forms. From the individual's
point of view, much more research into the biological and
psychological conditions of attitudes towards death is required before a definite answer can be given concerning any
invariant valuations of it. It is conceivable that a full and
qualitatively rich life, together with the inevitable biological
changes of aging, may carry along a more or less ready
acceptance of death as a terminus of life.[18]

To whom Is Life a Good? May we then conclude as a
basic moral judgment that in an unqualified sense life is a
good? Not entirely as yet. We can conclude that a man's
life is as such a good to him, that a people's or species' life
and its continuation is a good to it. And we can conclude
that one man's life is not as such an evil to another. But we
cannot as yet assert that the first man's life is recognized
by the second as a good to the first, nor that the first man's
life is, as such, a good to the second, rather than indifferent.
These depend on psychological judgments which we have
not yet considered, concerning the quality of interpersonal
relations, and on cultural and historical judgments concerning patterns of mutual good. That is why the fundamental
bio-psychic valuations we are now considering cannot be
thought of as fully articulated values; they are rather foun-

18. For a summary of evidence whether there is a desire for immortality,
see, for example, Corliss Lamont, *The Illusion of Immortality,* second edition (New York, Philosophical Library, 1950), ch. 6. For a different perspective, see Miguel De Unamuno, *The Tragic Sense of Life,* trans. J. E. C.
Flitch (London, Macmillan, 1931), esp. chs. III, VI. For the view that
every fear of death probably covers other unconscious ideas, see Fenichel,
op. cit., pp. 209 ff.

dations for value construction. Important as it is, to have
a continuity of life is only a beginning. It is good simply
to be alive, but man, unlike the amoeba, is also a social-
cultural and historical being, and his good is not thus, if
indeed it ever was in the past, a purely biological matter.

§ Evaluation of Pleasure and Pain

THE LOGIC OF THEIR JUSTIFICATION.   We turn from the
value of life to the evaluation of pleasure and pain. The logic
of the justification of pleasure and pain as good and evil
respectively, is in many respects parallel to that by which
life was found to be good. In rough outline it can be seen
in the arguments of Jeremy Bentham, the founder of the
Utilitarian outlook. At the opening of his *Principles of Morals
and Legislation,* after asserting that pursuit of pleasure and
avoidance of pain govern us in all our actions and in all our
moral judgments, he says: "In a word a man may pretend
to abjure their empire, but in reality he will remain subject
to it all the while." Hence he analyzed the ascetic view as
in fact a misunderstanding, miscalculation or distortion of
the search for pleasure. Philosophical partisans of asceticism
are really pursuing the pleasures of honor and reputation
by scorning pleasure. Religious partisans of ascetism are
motivated by "the fear of future punishment at the hands
of a splenetic and revengeful Deity." And some deprecate
pleasure because its pursuit seems to lead to pain; this is
"the reverie of certain hasty speculators," and is at bottom
simply pleasure theory misapplied.

The Utilitarians were, of course, concerned with a
stronger proposition than our present inquiry. They wished
to show that pleasure is the sole good and pain the sole
evil. We are dealing merely with the phase-rules "Pleasure
is a good" and "Pain is an evil," and asking to what extent
these may be regarded as true universal ethical statements.

SUBJECTIVE VIEW.   In modern times the predominant
ethical tendency has been to treat pleasure and pain as
subjective phenomena, immediately cognizable by the one

who is pleased or pained. They therefore require little description, since everyone really knows how he feels and what he feels. The result of setting up happiness as the human end seems therefore to be a relativistic individualism. Controversy centers about the measurability of pleasure, and whether there are describable qualitative differences that merit ethical rating.

ATTEMPTS TO EVALUATE "REALITY" AND QUALITY. The attempt to distinguish between real and unreal pleasures goes back to Plato. But in his case it seems to express chiefly a set of values. For real pleasures are characterized by their genuinely cumulative nature and their concern with the eternal (e.g., intellectual pleasures); the unreal come from the body's restoration to normal condition either from ill health or in the regular round of depletion and repletion.[19] Aristotle saw pleasure even in the latter context as positive and real. He pointed out that even in restoration of health it accompanies the creative effort of the healthy part, not the degenerative process of the diseased part. He characterized pleasure as the concomitant of unimpeded activity, and regarded happiness not as a feeling, but as a kind of structural quality of a whole good life.[20] With the subjectivization of pleasure in the dualistic philosophical tradition emphasis shifted to discernment of the inner quality. Thus J. S. Mill's attempt to distinguish qualities of pleasure rests simply on the appeal to the judgment of the person who has experienced the two classes of activities they accompany.[21] The career of pleasure in ethical theory has clearly been a checkered one, reflecting the current state of philosophy and of science.

BIOLOGICAL EVALUATION. The biological perspective goes beyond the mere contents of consciousness and seeks to re-

---

19. Plato, *Republic*, Bk. IX.
20. Aristotle, *Nichomachean Ethics*, Bk. X.
21. *Utilitarianism*, ch. 2. That Mill's distinction has wider social implications in his theory is suggested in my "Context and Content in the Theory of Ideas," in *A Philosophy for the Future*, ed. by Sellars, Farber and McGill (New York, Macmillan, 1949), pp. 433-435.

late the occurrence of pleasure to organic activities. Herbert Spencer, for example, looks directly to the role that pleasure and pain play in the evolutionary economy. His general thesis is that pleasure is on the whole correlated with beneficial actions (from the point of view of survival and expression of powers), while pain is on the whole correlated with actions that are injurious.[22] Since pleasure is a feeling we strive to bring into consciousness, and pain one we strive to remove, the species would scarcely survive if the correlation were reversed.

CONTEMPORARY PSYCHOLOGICAL ANALYSIS.   Contemporary psychology carries the study more minutely into the individual organism, gaining depth in a limited domain by the surrender of the wider historical canvas. Its procedures include both inspection of qualities in consciousness (which we shall call *phenomenal* inspection), the search for *causal* conditions, and the correlation of phenomenal qualities with underlying conditions. For example, Fenichel says: "the feelings of pleasure and pain as qualities are describable only; to 'explain' them means to determine under what dynamic and economic conditions they are experienced.

"This way of putting the problem would find a simple justification if a direct correlation could be found between fundamental quantities and the definite qualities that appeared only with them: for example, if Fechner's hypothesis —that every increase in mental tension is felt as displeasure and every decrease as pleasure—could be confirmed. Many facts are in accordance with such a viewpoint, but unfortunately there are contradictory facts, too. There are pleasurable tensions, like sexual excitement, and painful lack of tension, like boredom or feelings of emptiness. Nevertheless, Fechner's rule is valid in general. That sexual excitement and boredom are secondary complications can be demonstrated. The pleasure of sexual excitement, called forepleasure, turns immediately into displeasure if the hope of bringing about a discharge in subsequent end pleasure disappears;

---

22. *Principles of Ethics*, Vol. I, § 33.

the pleasure character of the forepleasure is tied up with a mental anticipation of the end pleasure. The displeasure of boredom turns out, on closer inspection, not to correspond to a lack of tension but rather to an excitement whose aim is unconscious."[23]

RESULTANT ETHICAL EVALUATION. Such detailed psychological analysis can help develop criteria for the ethical evaluation of different types of pleasures. Erich Fromm has attempted to do this in his *Man For Himself*.[24] His concentration is chiefly on separating the kinds of pleasure that reflect distortions of personality and basic insecurity (these have an insatiable anxiety-ridden quality) from the satisfactions that accompany relief from painful tension (hunger, thirst, etc.) and the joys that express a kind of psychological abundance (e.g., enjoyable taste experience as contrasted with food hunger, or sensual and emotional qualities of love as contrasted with sexual hunger).

The fruitfulness of such distinctions depends largely on the theory of personality underlying them. But their general effect is to untangle the various threads that in their confused state gave an "it depends" appearance to the goodness of pleasure and the evil of pain. The net result is largely comparable to that reached in the evaluation of life itself. The starting-point is the felt goodness of pleasure and evil of pain in consciousness. This is augmented by the widespread evidence of the pursuit of the one and the avoidance of the other. The over-all biological account correlating pleasure with activities beneficial to life and survival *adds* an instrumental positive evaluation. The similar warning or signal function of pain *subtracts* from its initial evil.

As in the case of killing, the vast infliction of pain by men on fellow-men in war and economic exploitation, is not coun-

23. *Op. cit.*, pp. 14-15. For a summary of experimental work, see J. G. Beebe-Center, *The Psychology of Pleasantness and Unpleasantness* (New York, D. Van Nostrand Co., 1932). For a careful phenomenological study of pleasure, see Karl Duncker, "On Pleasure, Emotion and Striving" (*Philosophy and Phenomenological Research*, I, 391-430, June 1941).

24. Erich Fromm, *Man For Himself* (New York, Rinehart, 1947), pp. 172-191, esp. 183 ff.

ter-evidence to the evil of pain. For the goal is not the infliction of pain, but the land or products, resources or services
obtained. In fact, historical evidence suggests that exploit-
ing groups seek to cover up in their own eyes, as well as
in the mind of the exploited, the fact that they are inflicting
misery, and they make the strangest ideological inventions
to achieve this purpose.

The association of pleasure with activities harmful to
the individual, as well as the actual pursuit of pains—both infliction of pain on others in sadistic phenomena, and self-
torment—carry us into the psychological problem of aggression discussed above. The fuller exploration of the impact
of psychological knowledge on evaluation is the topic of our
next chapter. But it will suffice to note here that the center
of gravity in such phenomena as alcoholism, drug addiction,
etc., has shifted in the light of increasing knowledge. It is no
longer a question of simply balancing in hedonistic fashion
the good of pleasure against the evil of consequences. For
the individual turns out to be concerned fundamentally with
the removal of anxieties, guilt-feelings and oppressive internal burdens, rather than with the pursuit of happiness.
Thus harmful pleasures are in the same broad class of phenomena as pursued pains. And in both self-aggression and
sadism—for Freud's picture of inward aggression turned outward may very well fit the pathological sadist without establishing a death-instinct as a normal component of all
men—the instrumental character of aggression stands out
clearly. Its aim is, again, to lighten the load of anxiety and
guilt-feeling.

Such lines of analysis and evidence would appear to
remove most of the obstacles to the general phase rules that
pleasure is good and pain is evil. But they do so only within
the same limitations as the goodness of life was established.
That is, we can assert that a man's pleasure is a good to
him and his pain is an evil to him. We cannot yet assert that
if one man is aware that another is pleased (pained), he
recognizes that other's pleasure (pain) as a good (evil) to
that other man. Nor can we assert the further step that if

the first is aware that the second is pleased (pained), the first regards the second's pleasure (pain) as a good (evil) in his own value reckonings. As in the case of life, psychological problems of human sympathy and its nature, and socio-historical problems of patterns of mutual good, remain to be considered before wider judgments may become possible. That our advance in valuation is so cautious does not mean definite indeterminacy. It simply means that there are further variables to be tracked down in their own specific fields. For example, could a culture so standardize infliction of pain (on others) that it would come to have a goal quality rather than a means quality?

§ Bases for Evaluating Biological Drives

EVALUATION PROBLEMS. Given the establishment as phase rules of the goodness of life and pleasure and the evil of pain, we may now go on to see whether a comparable determinateness can be achieved in the case of biological drives and impulses. We have seen that an "instinct" approach too readily assumed the goodness of instinctual expression in an ethical framework simply on the ground that instincts were constitutive of man. Rejecting this view does not, however, imply that there is no way of evaluating the expression of drives or impulses. Rather, it calls for a careful study of grounds for such judgments. We shall, accordingly, first look at shifting historical attitudes to human drives, then consider how a particular drive may be evaluated. We shall show that not merely the form of expression, but even the existence of a drive may be meaningfully evaluated, and that the judgment that a particular drive is on the whole evil is not ruled out. But the occurrence of such a judgment will not be found in conflict with a general phase rule that drive expression is a good, and considerable evidence will be seen to point in this direction.

SOME HISTORICAL ATTITUDES TO THE VALUE OF DRIVE EXPRESSION. Different attitudes to the moral role of biological drives is an old story. The conflict between Plato and

Aristotle on the question has been fairly typical of divergent views since then. Plato sees the impulses as blind, capricious, demanding—a many-headed hydra of desire which could be kept in line only by stern repression through will or spirit, at the command of reason.[25] Aristotle regards impulse, passion, feeling as all raw materials for the artistic work of character-formation.[26] None is condemned by itself as evil, and it is assumed that they will be given some expression. But evaluation is primarily of the resulting forms, for men want the good life, not simply life.

Historical Christianity follows now the Platonic, now the Aristotelian trend. It often tends to look upon drives as outside of the self, acting on the self, and so to limit evaluation to the attitude or reaction of the self to its drives. From this point of view impulse becomes a source of temptation, and its expression is suspect, if not definitely sinful.

Mechanistic views in reaction to the dualist and supernaturalist extrusion of impulse from the self, often stress the ultimacy of the biological drives, as if their constitutive role made their goodness inevitable. In similar fashion, considerable psychological writing takes the pleasure principle (that is, the tendency of a tension to seek immediate discharge) as basic, giving the reality principle (that is, the postponement of gratification in response to environmental demands) a secondary instrumental role. In both these outlooks, there is a complicated network of theoretical assumption underlying the transition from a constitutive or genetic account to a moral account. Central in this is the assumption that the individual not merely is that way or grows that way, but that in his mature activity these basic drives function as goals. In short, *all* activity *all* the time is really an attempt to satisfy these drives, and they are not to be seen only as sources of energy. It is precisely this assumption which has been questioned by psychologists who insist on the "functional autonomy" of acquired drives or aims, and by social scientists who stress the way in which developed

---

25. Plato, *Republic*, Book IV.
26. Aristotle, *Nicomachean Ethics*, Books III-V.

social values can revise biological valuations, as in denial of drive expression for social goals.[27]

A wider perspective is furnished by the evolutionary outlook, which carried the analysis outside of the individual and therefore ceases to regard drives as ultimate givens. Darwin saw instincts in this way as products of natural selection serving to further survival of the species.[28] They thus have the goodness that comes from playing a continuous fundamental survival and perpetuation role. Hence given continued life as a value, one is tempted to be thankful for hunger, thirst, sex, activity drives, as well as for any maternal and gregarious drives there may be.[29] A philosopher may even be pardoned for hoping that curiosity, so often listed among the instincts in the past, will turn out to hold a basic place along with food and sex. And there are many other qualities one would indeed be glad to have had implanted as drives.

ON EVALUATING PARTICULAR DRIVES.   Such an evolutionary perspective is sometimes taken to justify all existent drives. Their very existence is taken to show that they developed by furthering human survival. But even if this were historically true, it does not follow that their present character is the same. A drive might have had a useful role in the past, and yet have a disruptive effect under changed conditions. It may be judged evil because of the disruption, and be judged to have lost any goodness it once possessed because of the change in conditions and circumstances in which it had application. Thus if aggression is instinctive, we can see how it may have suited an environment of continuous danger, and still be quite out of place in a complex industrial society. But if it is instinctive, it must be found

---

27. This issue will be discussed below in connection with the theory of needs; see pp. 167-171.

28. *The Origin of Species*, ch. 8.

29. Julian Huxley points out, for example (*Touchstone for Ethics*, p. 134), that sexual fusion—as contrasted with merely division of the old individual into two new ones—made possible the pooling of individual mutations. Subsequently reason made possible pooling of individual experiences.

some mode of expression. Hence William James looks to a moral equivalent of war in socially useful energetic collective undertakings by the youth, and others look to sport. Similarly, anger as an emotional reaction may well have had a useful part to play in survival and in aggression readiness. But what shall be said of it ethically? The Sermon on the Mount almost equates it with murder, whereas Nietzsche insists on allowing it even towards friends because one wants to keep in trim against one's potential enemies.

It follows that the mere existence of a drive does not determine whether it is to be judged on the whole good, bad or indifferent. But its evaluation in terms of qualities and effects is not therefore an arbitrary one. In any case, if drives are more than neutral energy, their specific forms of expression will also require further evaluation. The criteria so far established, in terms of life, pleasure and pain, will always be applicable. Psychological evaluation may go on to distinguish normal from disguised expression. For example, the characterization of some actions as perversions, with a negative value connotation, appears to rest on the view that drive expression has been blocked and is incomplete.[30] On the other hand, when some actions are taken to be sublimations, the assumption is that the drive has found expression which will have no disruptive effect on the person, although the goal object is not the one with which that drive is directly associated.[31] Similarly, forms of expression may be evaluated from a cultural and historical perspective as well, in terms of their specific relation to social forms which they promote or hinder.

POSSIBILITIES OF DISCORDANT ELEMENTS WITHIN THE BIOLOGICAL NATURE OF MAN. One very important consequence of the evolutionary outlook in this application is that it leaves open for scientific determination the issue whether man has a unified or internally discordant biological nature.

---

30. See below p. 185.

31. See below p. 184. A complex theoretical network clearly underlies such conceptions. Care must be exercised in any context to see that such judgments do not represent arbitrary valuations.

There can be no a priori assumption of systematic unity other than that required for bare survival. It is quite possible that in man the long development of the organism should have produced some invariable frictions which endured because in other respects the specific features aided survival of the species. Julian Huxley somewhere calls attention to the difficulties that followed man's assumption of an erect posture. Roheim suggests as a basis for the psychological problems of the child the existence of conflicting biological factors—relative precocity in sexual development combined with retardation of bodily development in man as compared to lower animal forms.[32] Some fundamental moral issues may therefore have to be traced all the way to conflicts in "human nature."

BASES FOR A POSSIBLE PHASE RULE THAT DRIVE EXPRESSION IS A GOOD.   In spite of the fact that a particular drive might be judged evil, and that discordance is possible within human nature, it is still possible to consider a phase rule that drive expression in general is a good. This does not imply that it is always wrong to hinder or postpone expression, but that in assessing a situation expression is always to be reckoned a good, even where it is denied in determining action because other weightier values point in the opposite direction.

On the whole, a growingly favorable judgment on the value of drive expression does characterize contemporary attitudes. This may stem to a great extent from the growth of secular naturalist outlooks and their reaction to previous opposing views. But there is some justification in the lessons of experience and scientific inquiry, although the issue is not yet so sharply formulated nor the evidence so decisive as to yield a settled conviction.

There is in the first place the recognition that men's drives press for fulfillment. The tension that arises is basically painful, and the reduction of tension a source of great

---

32. Géza Roheim, *Psychoanalysis and Anthropology* (New York, International Universities Press, 1950), ch. 10, esp. pp. 402 ff.

pleasure. Even any adjudged "bad" drives require some accommodation. The phase rule is thus supported by judgments of the goodness of pleasure and the evil of pain. On the other side, however, is the recognition of the many ways in which an inability to tolerate tensions may be itself a pathological phenomenon. Goldstein suggests that the normal situation is rather "the *formation* of a certain level of tension, namely, that which makes possible further ordered activity."[33] But such an analysis raises in turn the question whether intolerance of tensions does not itself reflect some degree of blocking of drive expression; intolerance of tension would then be a secondary phenomenon within a framework assuming the goodness of drive expression as a phase rule.

There is in the second place, a greater awareness of the direct relation of positive values in human character (for example, the ability to have sympathetic interpersonal relations) to appropriate development of basic impulse expression. Hence drive expression has an intimate relation to the sources of moral virtue.

Moreover, repression as such has lost the ethical nobility it possessed when it was regarded as issuing from reason in the battle of reason versus impulse. There is too great a sensitivity to the wide psychological evidence that intense repression often proceeds from the same drives themselves taking a distorted form and expressing themselves in the act of repression.

There are, again, the multiple instrumental and raw material functions of biological drives—both the general survival role they play on the whole, and the specific energizing functions in relation to many particular physical, cultural and social goods. To take a simple example, the admitted joys of social festivity centering about eating require hunger as raw material. On the other side, however, are to be reckoned the occasionally disturbing social effects of some drives. As

---

33. Kurt Goldstein, *The Organism* (New York, etc., American Book Co., 1939), pp. 195-196.

against this, and further supporting the general phase rule, are the lessons of experience that impulse disturbance is often due in great measure to unfavorable social and cultural arrangements. Hence when wide social disturbances do result from insufficiently controlled drive expression (as in problems of sex in relation to delinquency) there is now less tendency to rush in attack on the impulse than to look to defects in the social and cultural patterning.

The reaction against the traditional restrictive attitude to impulse expression whose evils have become all too clear is no mere swing of the pendulum. In addition to the psychological knowledge it embodies, there are social factors into which greater insight is now possible, and firmly established social attitudes, both of which add to the basis of justification. The central insight is that a major prop in a restrictive attitude was the objective scarcity of means for the widespread satisfaction of human impulse. With the increase of material means this prop has been removed.[34] The major social attitude is the liberalism referred to above,[35] which incorporates a strong individualism and is *for* any impulse anywhere, provided it does no harm to others.

Some of the strands in an over-all conclusion require amplification from the detailed points of view of the other sciences, and some, no doubt, the growth of human knowledge may clarify further. But in any case, the evidence is sufficient to reject a general negative valuation of drive expression, and even to incline to the positive rather than merely neutral view.

§ Evaluation of Health and Frustration

THE IDEAL OF HEALTH. If we combine the results we have found concerning life, pain and pleasure with the proposition that on the whole expression of drives or impulse is a good, two very important corollaries would seem

---

34. The role of increased abundance will be considered more fully below, pp. 320 ff.
35. P. 61.

to follow: one, that health is a good, the second that frustration is probably an evil.

The ideal of health is obviously a synoptic one, referring to good condition of the organism in many different respects. It therefore covers appropriate functioning of parts, adequate expression of drives, absence of persistent pain, capacity for pleasure, and so forth. The value elements in the ideal are therefore the ones already considered; no fresh value is smuggled in under the notion of "appropriate" functioning or "good" condition. The prominence that the ideal of bodily health has come to have in parts of the world such as the United States no doubt reflects psychological and historical components as well as biological, but basically, it can now be agreed, everybody wants to be healthy. The treatment of evidence that appears to conflict with this is parallel to that in the case of life, pleasure and pain. It yields the minimal conclusion that health is not an evil, and that it is an instrumental good, and the probable one that it is an actual good.

THE MEANING OF FRUSTRATION. Frustration is commonly thought of as conflict that prevents action already on the way or strongly desired. But if the judgment that frustration is evil is not to be a cursory assumption, its meaning and the grounds for it must be more deeply explored. For example, it might seem that over and above the evil of pain and the good of pleasure and expression of impulse, there is a positive valuation of action, and any such elements should not be left implicit.

The need for action, and the primary role of action in human life can scarcely be doubted. But "action" is a weasel word. If I am hit on one cheek and strike back, that is action. But if I turn the other cheek, is that not also action? And if I do nothing but simply curb my resentment, is that action? And finally, if I do nothing, curb nothing, but simply watch the flow of my feelings, is that different in type? There are no doubt distinctions worth making here about the quality of action, but they rest, as we shall see, on psychological assumptions, not on definitions.

What is frustration itself? To explore this look, for example, at the "Experimental Studies of Frustration in Young Children" by Barker, Dembo, Lewin and Wright.[36] They point out that "Frustration occurs when an episode of behavior is interrupted before its completion, i.e., before the goal appropriate to the motivating state of the individual is reached. It is well established that frustration has widely ramifying effects upon behavior" and also that "To frustration has been attributed most of what is valued and deplored in individual and group behavior: delinquency, neurosis, war, art, character, religion. This, however, is speculation. It is of greatest importance that these questions be removed from the realm of speculation, that the conditions of frustration be conceptualized, that its degree and effects be measured and that systematic experimental studies be made."[37] This introduction alone makes the theoretical point with which we are here concerned. For if opposite claims are made for frustration, if goods and evils are assigned to its workings, if, in short, it can be applied wisely, how can it be condemned in advance? A decision is required in the light of evidence, rather than a simple deduction from the evil of pain and the good of pleasure.

EXPERIMENTAL EFFECTS OF FRUSTRATION. Let us, however, press the inquiry further to determine more carefully what frustration is and does. The experimenters allowed children to play with attractive toys, then by use of a wire-net barrier separated the children from the toys and observed what they then did. Less attractive toys were still available to them, but the more attractive remained inaccessibly in sight. The measure of the strength of frustration was "the proportion of the total experimental period occupied by barrier and escape behavior."[38] The results pointed to a regression in the level of intellectual functioning. Also "there was a marked decrease in the happiness of the mood (e.g., less

---

36. Summarized in *Readings in Social Psychology,* ed. Newcomb, Hartley and others (New York, Holt, 1947), pp. 283-290.
37. *Ibid.,* p. 283.
38. *Ibid.,* p. 287.

laughing, smiling and gleeful singing), there was an increase in motor restlessness and hypertension (e.g., more loud singing and talking, restless actions, stuttering, and thumb sucking); and there was an increase in aggressiveness (e.g., more hitting, kicking, breaking, and destroying)."[39] There was also an increased intra-group unity and increased out-group aggression; this the experimenters are also inclined to interpret as regression.

VALUES INHERENT IN THE EVALUATION. If we judge frustration to be evil on the basis of such a study as this, the assumed values are clear enough. A more advanced intellectual functioning in those capable of it is better than a less advanced one. Happiness is better than unhappiness. Restlessness, hypertension, including stuttering and thumb sucking, are undesirable. Aggressiveness and destructiveness have negative value. Intra-group unity that expresses regression has an element that diminishes the value of the unity as such. There are very definite values involved here. Some may be subsumed under the good of pleasure and the evil of pain; some under the failure of expression of impulse. Regression is more complicated. It can not be considered bad merely because it implies an older person using mechanisms of response more characteristic of an earlier stage. To judge this bad must involve some theory of development which determines a value framework. And this is a further psychological question.[40]

FURTHER CONSIDERATIONS. It is worth noting furthermore that even such results as are found in this experiment are not conclusive for all frustration. The outcome of the experiment is a function the specific conditions of the field. As Gardner Murphy points out, "Had the children

---

39. *Ibid.*, p. 288.

40. See the discussion of maturity below, pp. 179 ff. There are also such issues as whether the results of the experiments can be explained in terms of alternative theoretical approaches. See, for example, Irvin L. Child and Ian K. Waterhouse, "Frustration and the Quality of Performance: I. A Critique of the Barker, Dembo, and Lewin Experiment," *Psychological Review*, 59, 351-362 (September 1952). For selected other materials and analyses in the general field, see Newcomb and Hartley, *op. cit.*, Part VI.

been led into the playground, they might well have run
faster, kicked the football harder, gone down the slide more
courageously, and done many other things which are *less
aggressive,* also less *regressive* and more 'mature,' i.e. corresponding to a developmental level which they had not yet
reached."[41]

In a later context, however, Murphy says of the term
"frustration": "The word is useful in indicating not merely
the attainment of a state of suspense but the irradiation
through the organism of a high tension, that is, a suffusion
of stress to other parts than the one originally involved. It
is important to emphasize clearly that this spread of tension
is no metaphor. The tension can be followed step by step
with Jacobson's technique for measuring the magnitude of
muscle tensions electrically; waves of increasing or decreas-
ing tension can be followed all the way from fingers and
arms to the gastrointestinal canal. Stomach ulcers can be
directly treated by re-educating the innervation of arm and
trunk muscles. Frustration, then, is not only a negative thing,
a failure to achieve; it is a positive augmentation of tension
throughout the organic system."[42] The temporary consequences of frustration are outbursts of poorly conditioned
responses; and there may be permanent consequences in
defensive expressions of posture. The fact of break-down
under frustration is, of course, real enough, and it involves
the cessation of the ability to function. The classical laboratory case is Pavlov's dog which, conditioned to opposite
reactions by clearly differentiated stimuli, was then given
an ambiguous stimulus—e.g., to salivate when shown a circular spot of light, to inhibit salivation when shown an ellip-
tical spot, and then presented with a broadened ellipse close
to a circle. It must, of course, be remembered that the dog's
possible reactions were limited; it was not, for example, free
to run away. The result was a "break-down" phenomenon.

RESULTANT EVALUATION OF FRUSTRATION.   Frustration

---

41. *Personality, a Biosocial Approach to Origins and Structure* (New
York, Harper, 1947), p. 146.
42. *Ibid.,* p. 305.

as a human phenomenon clearly requires fuller identification and analysis. How far it is to be regarded as a single kind of phenomenon with a general component, how far it simply embraces analogous specific types, are questions requiring further research. A great part of this work involves psychological perspectives. If frustration is to be judged an evil within an ethical framework, as we should be inclined to regard it at this stage of knowledge, it is not merely because of the bio-psychic values already considered but also and perhaps decisively because of its negative place in the more systematic picture of personality development and human needs to be considered in the next chapter.

It should be added again, by way of warning, that the evaluation of frustration has the same unsettled aspect in social relations that life and pain had. The admitted evil to the individual of frustration and breakdown does not itself settle the question of social morality concerning the distribution of the evil. Just as men for a long time apparently believed that prostitution, an admitted evil to the women concerned, was yet a safeguard against the violation of marriage and its sanctity, so the breakdown of many has seemed not too high a price for the maintenance of conditions which generated the frustration—at least to those whose happiness depended on these conditions. This is, of course, a callous view. But we may not, in such an investigation as the present, smuggle in our democratic equalitarianism by way of definition or assumption. If it is to enter, let it come by way of the front door of investigation, with its head high.

§ The Functional Hypothesis

ITS MEANING. We turn from specific valuations for which the biological perspective provides a foundation, to its methodological contributions in analyzing ethical relativity. Of these the most important is its functional hypothesis, developed especially in the evolutionary outlook. The functional concept has been widely used and widely criti-

cized in the psychological and social sciences, and some
consideration of its issues will be found in subsequent chapters. But especially since it is relatively simpler in the biological context, we may begin its analysis at this point.

A set of phenomena have a functional role if they play
a part in some process or enterprise beyond themselves
which they either make possible or more likely to be successful by their participation. This implies, of course, that
the character of this process or enterprise is antecedently
determined and the meaning of "success" likewise understood. The relative simplicity of the concept in biological
usage comes probably from the established character of the
goal of *survival*. Complications arise chiefly where it is an
issue of individual versus group survival; Leslie Stephen
points out, for example, that the promptness with which the
female sacrifices herself for offspring has group, not individual, survival value.[43] Other complications stem from a
confusion of causal investigation and description. Thus to
find a phenomenon playing a role does not imply that it
came into existence to perform this role nor even that it
continues in existence to do so. These are separable questions. Automobiles came into existence to perform a role;
sunshine did not do so nor does it continue in existence for
agricultural purposes; rain comes into existence to carry out
a role only when the clouds are seeded.

The functional hypothesis in relation to moral and ethical phenomena is that they always serve some functional
role in human living. It does not state whether the role is
primarily biological, psychological, social, historical. But
the simplest evolutionary sketch of man living in groups
struggling for survival raises the possibility that men's moral
feelings and judgments have served some survival role—
for example, maintaining a cohesiveness of the group without which survival would have been difficult.

ITS SCOPE. Such a principle may be applied throughout

43. Leslie Stephen, *The Science of Ethics* (New York, Putnam's, 1882),
p. 91.

the whole range of moral qualities and ethical processes, from simple valuing to complicated analysis. In the case of drives or needs, that for food, for example, appears in consciousness as the attempt to assuage the tension of hunger feeling. But getting hungry itself is an important ability for an organism that requires replenishment in order to continue living. Similarly, pain has the functional role of danger signal (although, unfortunately, it does not always work), and fear has a comparable role. Imagine how safe an environment would have to be to dispense with fear as a mechanism, and how healthy an organism to be able to discard pain.

In dealing with obligation feelings, Julian Huxley compares our ethical mechanism to our backbone as a supporting mechanism. The ethical mechanism operates to charge whatever passes through it with emotive qualities of rightness and wrongness,[44] thereby serving psychological functions for the individual by providing a kind of security until his reason develops, and cultural functions for the group as a means of transmitting and sanctioning group mores.

Evaluation and ethical theorizing may likewise have a functional role. And the role itself may vary at different times in historical development. To take another example from Julian Huxley's treatment, he finds ethics in pre-agricultural communities to be largely concerned with group solidarity, in early civilization more especially with class domination and group rivalry. And he looks for an evolution in the ethical development as part of the evolutionary process.[45]

The hypothesis that moral and ethical phenomena have functional roles, these examples show, extends through all the sciences, and is not limited to the biological perspective which gives it its broadest foundations. Moreover, the analysis implies that we have not merely different types side by side, but that all interact in possibly complicated structures.

---

44. T. H. Huxley and Julian Huxley, *Touchstone for Ethics*, pp. 116-117. See above, p. 39.
45. *Ibid.*, pp. 127, 131 f.

We shall meet these features again in psychology, the social sciences, and history.

IMPACT ON THE ETHICAL RELATIVITY PROBLEM. The general effect of a functional hypothesis is to narrow down indeterminacy in ethical judgment. If aspirations and strivings and theorizings are part and parcel of the concrete processes of human living, and play definite roles, then evaluation becomes increasingly possible in terms of the successful carrying out of the specific roles. The factual question is whether there are actually such roles and whether they have a definite structure. And in the broad sense in which biological evolution shows a process of effort for survival it does lay the groundwork, as we have seen in the specific discussion of the bio-psychic values, for an affirmative answer.

It is important, however, to underscore the structure of the process by which any valuational indeterminacy with respect to the roles themselves is dealt with. It is obvious that in general the development of a mechanism which serves a functional role does not automatically establish the value of the role, nor that the mechanism ought to continue to be used for that role, if it is possible to divert it or ignore it. Suppose, for example, it is true that morality developed as part of the effort of social groups to hold together in order to survive. One could ask whether it is still equally important, whether morality itself as cohesiveness is required to the same degree now that man has stabilized his position. Perhaps it is part of our lowly past. Even if the role of the mechanism is inevitable, one could ask why it should not be conceived as a necessary evil rather than as a good. One could even ask, as we have seen in considerable detail, whether survival and life itself are good. But one should not assume that just because it may be asked therefore it cannot be answered.

In the light of the misuse of the biological perspective in the last century, it is important to see what is involved in this continual demand for further evaluation, and what the functional hypothesis does not assert. A functional hypothesis does not assume that functions may not undergo change: historical functions certainly do and they may give form and direction to biological and psychological functions. Similarly, there is no invocation of "adjustment" to existent trends and forces as an ethical criterion. An individual may always reasonably ask why he should adjust to a given condition of society rather than have it adjust to his needs and criticisms. Nor can we say with simple optimism that whatever is coming to be must be for the best. At every point, once evaluating has arisen as an ethical process within human life, its right to question the value of any existent form, trend, value, theory, or anything else, is part of its very being as an ethical process.

But there is one thing that the questioner engaging in the evaluating process will not be permitted to do by a mature evolutionary ethics. That is to step outside of the evolutionary process altogether. For that would imply that his standards of good and evil, his aims and strivings, his own concepts and theoretical frameworks are not themselves *inside the evolutionary process*. That would imply that some part of the phenomena by stipulation is exempt from naturalization and immune to the search for functional relations.

THE BATTLE OVER TRANSCENDING THE EVOLUTIONARY PROCESS. This brings us to the philosophical heart of our problem—the attempt to separate ethics from the totality of natural and human relations and give it an external or transcendent standpoint. T. H. Huxley became a spokesman for a position with such implications in his Romanes Lecture of 1893, "Evolution and Ethics." He had himself done much to stereotype the tooth and claw or gladiatorial conception of evolution. Now he simply cut the knot between the cosmic evolutionary process and the human ethical process. He saw the latter pitted against the former as a well-tilled garden in a wild jungle that continually encroaches upon it. Sensitive to the multitude of forms in which dualism insinuates itself into our thinking, John Dewey, shortly after Huxley's lecture pointed out that the phenomena did not require such a reaction.[46] It was not the ethical process that was opposed to the cosmic process, but one part of the cosmic process to another part. Dewey takes Huxley to be saying that man is an organ of the cosmic process in effecting its own process. That nature is always trying to reclaim what man has conquered shows simply that environment minus man is not the same environment as one including man. And he concludes: "But I question whether the spiritual life does not get its surest and most ample guarantees when it is learned that the laws and conditions of righteousness are implicated in the working processes of the universe; when it is found that man in his conscious struggles, in his doubts, temptations, and defeats, in his aspirations and successes, is moved on and buoyed up by the forces which have developed nature, and that in this moral struggle he acts not as a mere individual but as an organ in maintaining and carrying forward the universal process."

Julian Huxley, in his Romanes Lecture "Evolutionary Ethics" fifty years after his grandfather's—the shift in title is significant—makes the same point as Dewey, in a more activist vein. He reminds us, in effect, that the garden and the gardener are themselves evolutionary products, and that with the tremendous growth of human knowledge they play an increasingly greater role in determining the path of evolution.

The strength of the more sophisticated evolutionary theories has never come simply from a naive identification of the good with what is evolving. It has rested rather on the way in which the development of standards of judgment, of aims and strivings, of concepts and theories, has been incorporated into the process itself. There is no outside observer, no reflective inquirer estimating the whole process. Differences of evaluation are rather different strands within the process, representing conflicting elements. For example, the criticism of the group cohesive function of morality men-

46. John Dewey, "Evolution and Ethics," *The Monist*, VIII (1898), pp. 321-341.

tioned above may represent a later stage of stabilized human
domination in the animal world in which morality is capable
of being used to develop diversity instead of conformity.
Or it may represent a self-assertiveness itself developed in
the evolutionary process in carrying out some function.
Thomas Huxley believed that such self-assertiveness was
an inheritance "from the long series of ancestors, human and
semi-human and brutal, in whom the strength of this innate
tendency to self-assertion was the condition of victory in
the struggle for existence."[47] A contemporary analysis would
stress cultural and historical patterns more in the explanation
of self-assertiveness, but in principle the theoretical type of
explanation is the same. In either case, self-assertiveness is
not evaluated by being abstracted from the process and
held up to inspection by an outside intelligence which nods
its "good" or "bad." It is rather appraised for the way in
which it meets the problems to which it is addressed, the
criteria being embedded in the structure of the problems
at that point in the evolutionary process. Values, even as
standards, are thus not permitted the luxury of transcending
existential processes.

The extreme form of the anti-functional view is seen in
attempts not merely to have values transcend existence but
to have man transcend all values. For example, Eliseo Vivas,
discussing the passage of a man from what he calls the
moral life to the ethical life, in which a man overthrows his
dominant values in a conversion-like experience, declares:
"when we repudiate our constitutive values altogether and
forge an entirely new personality, a naked, empty self must
do the choosing." However, he himself adds the comment,
"and how this process is possible is not at all obvious."[48]
Similarly, there is a tendency among many ethical theorists
to speak of values as *commitments* of the self in such a way
that it looks as if one could separate all values from the
self and then inquire why one should reattach them.

---

47. In "Evolution and Ethics," *Touchstone for Ethics,* p. 53.
48. *The Moral Life and the Ethical Life* (Chicago, University of Chicago Press, 1950), pp. 230-31.

But not even this last refuge is permitted by the functional hypothesis. The self itself is within the evolutionary process. A naked self would be not a something outside acting on transcendent principle, but a strange dead-end product, a being denuded of all values, even—if we carry it logically to an extreme—of the desire to have values. How could such a naked self choose anything? As an empirical hypothesis then it may be suggested that the apparent phenomenon to which Vivas points is only a surface appearance. The biologist will insist that the human organism remains clad with some fundamental needs and drives, whether they appear on the surface or not. For the psychologist, the feeling of absolute novelty in conversion experiences comes from ignoring the accumulating value-pressures which precipitated them. Similarly, as the social scientist and historian see it, the sense of absolute novelty in a social revolution with sharp revision in social forms comes largely from ignoring the accumulation of social needs and grievances which were hammered into a revised value-outlook. Yet a naked self choosing is not a logical contradiction. It is simply that the trend of evidence is against it. And this trend has reached the point where even *definitions* of the self are beginning to be cast in terms of values.[49]

In general, we may conclude, the functional hypothesis leads us to probe continually beyond the initial frame of the presented phenomena, for extended role relationships. It constitutes a major theme in an evolutionary outlook. It is not to be applied mechanically, for it does not of itself tell us the closeness, intensity or extent of the ties sought, nor within which scientific area they fall. The discovery of functional role does not replace whatever intrinsic qualities or values a set of phenomena may exhibit. It may strengthen them within the wider evolutionary perspective, or occasionally work against them.

---

49. For the study of values and the self, see below pp. 176 ff.

§ Group Orientation within the Biological Perspective

EFFECT OF LONG-TIME BACKGROUND.   The second analytical tool developed in the biological perspective is an inclination to examine questions of value as group phenomena. The biological perspective does not make a decisive choice between the group and the individual. There is rather a constant reminder that we must not stop with the individual as final. The constant background is a long-time one. It is the movement of the whole evolutionary process, with the efforts of populations to survive and extend themselves. It is true that when we ask why survival is worth while, why the group should maintain itself, we are sent to the biological strivings of individuals. But this is because life itself is a property of individual organisms, and drives are embedded in individual structure. We are not referred to any particular individual, nor to the search for personal satisfaction. Instead, the search is for invariant or structural elements in the strivings of men.

The group orientation appears again when we seek to evaluate the expression of biological impulse as well as specific drives. For then we are sent back to evolutionary considerations and the role of drives in survival. Similarly, phenomena of emotions, pleasures, pains, are all considered not so much for their qualitative character in consciousness as whether they are correlated with biological processes that are beneficial or harmful for individual and group survival.

THESIS THAT MORAL PATTERNS ARE GROUP PHENOMENA. Such a group orientation suggests the fundamental thesis that moral patterns are group phenomena, that they have always been in some sense operative in relation to the group. This insight of the biological perspective is, like its other major insights, amply confirmed by the social and historical sciences. Every actual morality is found on investigation to be serving some societal or group role. This does not preclude an individual inventing his own morality with his own criteria; but it suggests that such rare occurrences are better understood as reaction responses to specific problems in some group morality. It is the same as with a language. The various "natural" languages, as the logicians call them to distinguish them from conventional logical or mathematical languages, are cultural-historical products in the manifold communication processes of persons. The fact that individuals construct conventional languages for specific scientific-cultural purposes does not therefore mean that a scientific linguist has to analyze a language as the coincidental sound preferences of individuals making isolated choices—even if such an analysis is theoretically possible. Similarly, the fact that some individuals propound individualistic moralities does not mean that morality has to be analyzed as individual assertion of personal will. An individualistic morality can itself be seen in group perspective, as a proposal for relaxed as against stricter cohesiveness, to be justified by its fruits for the group, just as a given system of private property may claim to be socially desirable, rather than just to rest on absolute individual rights.

Now such a group formulation can easily be read as an arbitrary value-preference for social-mindedness. The group orientation of the biological perspective suggests, however, that over and above the value of social-mindedness there is a scientific sense in which morality is a group phenomenon. In such a perspective the central problem for the individual, in contrasting himself with the group, would not be "Why should I yield to group demands?" but "Why should I *become* an island apart?" Obviously this problem points to a fuller reckoning, psychological and social as well as biological, with the phenomenon of egoism.

§ How Far Can the Biological Perspective
   Reduce Indeterminacy in Ethical Judgment?

SUMMARY. We have seen that the basic contributions of the biological perspective are foundations, not complete moral patterns. The naturalization of man opens his ways to empirical study and yields a basic picture of structure and development. The functional hypothesis forges an analytical tool, and the group orientation guarantees its wider use. The several bio-psychic values established constitute phase rules for application in particular judgment. To this extent we have some basis for systematic ethical knowledge. It is quite likely that more extended investigation with reference to ethical results would yield further phase rules to add to those concerning life, pleasure, pain, biological drive expression, health, frustration. For example, there may be sufficient biological evidence to warrant a judgment concerning the general value of cooperation or feelings of sympathy, along the lines sought by Kropotkin; or concerning integrative properties of systems along the lines followed in the homeostatic concepts. But the psychological and social content in such inquiries is likely to be even greater than in the basic valuations examined above, so that it is doubtful whether the evaluation could be fruitfully carried out in limited biological terms.

PARTICULARIZATION OF THE PHASE RULES ESTABLISHED. The bio-psychic values dealt with are capable of indefinite expansion and clarification in terms of the progressive results of the biological sciences. This is obvious in the biological ideal of health, if we look at it in the history of medical science. Similarly, the ideal of extended individual life will be increasingly clarified as we learn what is controllable and what is unavoidable in the loss of physical and mental powers in old age. Many moral decisions are now perfectly clear compared to what they used to be, and many more may be much further refined. The whole ethics of pure food, water, milk, drugs, as well as safety with respect to fire, vehicle accidents, air purity, overcrowding, constitutes a system of relatively stable value judgments. These do not cease to be ethical in character simply because they are thoroughly cemented with factual information. An adequate diet is not an external means to health, but a part of the process of continued healthy living.

Particularization will also serve to bring into systematic

unity those bio-psychic values that may seem, because of external conditions, to come in conflict with one another. For example, from the point of view of continuity of life alone it may seem to be less important whether the present generation is healthy than whether the whole population keeps on increasing. Individual life expectancy may be under thirty, but each man may leave a far greater number to succeed him as he goes by on the stage. On the other hand, a greater stress on health in each generation may require a diminished or at least stabilized population in many areas. However, conditions requisite for both values may bring them closer together. Basic attention to health is required to prevent epidemics that might jeopardize continuity— even apart from further complications on the psychological and social level characteristic of a submerged and marginal life. The result with the growth of a consciousness of the problem is a unity of the apparently conflicting values in the ideal of an "optimum" population for a given area in the light of available resources. Issues involving conflict in such an ideal no longer stem from bio-psychic valuations but from judgments of available resources (which may entail reckoning in political terms with other countries) and judgments of desirable standard of living (which may entail direct evaluation of larger families as compared with greater comfort for the smaller). Here again, the relevance of psychological and social perspectives to particularization and application becomes clear. But it must be remembered that even if these should turn out to have large elements of indeterminacy the latter is occurring *within* a framework of already established values, hence within a partially determinate value field.

CIRCUMSTANCES GIVING BIO-PSYCHIC VALUES A DECISIVE ROLE. On the other hand there are some conditions or states of the globe which, in a given age, may make even biological values decisive in resolving moral problems. Suppose, as indeed seems to be the case in the light of available atomic and biological means of warfare, that the probable outcome of a next war is either the wiping out of most human life on the globe or at least the general destruction of health. Does not then, the sheer value of simple life and health become basic and determine a moral decision of the necessity of peace? Even the argument that there are some things more precious than life and health would not prevail. For this concerns individuals or groups taken singly, not the whole human race without whose existence nothing more precious than life could even have a chance of future existence. Thus, given the analysis of the causes of war in the modern world it follows that the necessity of peace provides a basis for a moral critique of any human institutions, social practices, character-traits, that stand in the way of keeping the peace. Anything from national pride adding to friction to private ownership of national resources that require development might conceivably come under the moral ban.

Nevertheless, although in the extremes in which we are tending to find ourselves, a purely survival ethics might come to serve as the basis for a determinate common-human ethic, as it probably did in the very early past of mankind, it is to be hoped that something less desperate will be the case.

VI. Psychological
Perspectives

If we look back to the account of the threads woven into the fabric of ethical relativity, described in Chapter I, we can see that at least half of them are made of psychological materials. There is the egoistic view of man's nature, the assumption of a drive for power, the mechanistic belief that the mind is like a wax tablet on which he who wields the stylus may fashion the pattern. In addition, a great deal of the assurance of the stubborn man comes from the confidence that his own consciousness is a court of last resort in ethical judgment. The fate of an arbitrary ethical relativity is thus closely tied to the lessons of a science which has taken the nature of man as its province. In this chapter we shall not consider all the contributions that psychology has made or can make to ethics. We shall concentrate solely on the contributions that impinge on the problem of relativity. For there are a number of points in current psychological theory which suggest that we may say, "There is a definite answer—even if we do not yet have it," rather than simply, "It depends . . ."

An ethical theory that rested on the older psychological outlook has to revise itself in the light of the newer lessons of psychology. It has to learn that introspective data on values, while definitely data, are not indubitable last words. It has to recognize, as against the completely plastic "anything goes" conception of man, that there are fundamental human needs probably of an affiliative sort, which may be ignored only at the price of distortion. It cannot ignore the accumulation of evidence about the pathological forms of power pursuit, nor the effect on traditional egoism of psychological study of the self. It has to reckon with the impact of a growing theory of personality development which may make possible scientific ethical judgments of maturity and immaturity, and not simply as subjective biases in the scientist. It must face the growing body of materials on the phenomenon of amorality, and the contemporary sophisticated search for value invariants.

These are the possible avenues of diminishing ethical indeterminacy that we shall explore. But it is important to recognize at the outset that they are directions for further research rather than sealed pages of knowledge. Psychology herself is in the throes of rapid development and is grappling in many forms with the theoretical issue of her own individual-centeredness. We shall have to look also at the fundamental limitations which this individualistic orientation has imposed on the psychological perspective.

§ Consciousness and Values

CONSCIOUS REPRESENTATION OF ONE'S VALUES NOT INFALLIBLE. A first and very important lesson of contemporary psychology—one noted in our encounter with the stubborn man—is that a person's consciousness does not necessarily represent his real values accurately. Psychology provides some techniques for probing critically into the nature of a person's preferences. Such an approach goes beyond external criticism of a man's aims, such as that they are unattractive to others, or harmful to others or even in the long run to the man himself. It challenges the aims or preferences on home ground. That people "fool themselves" about what they want is a homely lesson of ordinary experience. The systematic uncovering of the manifold distortions in consciousness is associated with the Freudian development of the theory of unconscious motivations.[1] Even though there are numerous disagreements among the various modern schools of psychology, they agree predominantly that consciousness can give an incomplete or distorted view of a situation.

It is by now a familiar enough fact that a particular goal to which an individual clings may turn out to have a "spurious" quality, a particular character-trait he prizes as a virtue may in reality play a merely instrumental role in his personal economy, that he may be unaware of his actual feelings or misinterpret their nature. For example, a highly intelligent and capable young man persists in aiming at becoming a lawyer but something "accidental" goes wrong at every examination. On analysis it turns out that his adherence to the aim does not express a positive desire to be a lawyer—he surrenders this aim with relief—but simply unreasoning obedience to the wishes of a stern father no longer living. Again, strong ambition may, in a particular case, express a driving insecurity rather than a striving for a genuinely adopted goal. A character-trait one may exalt, such as great solicitude or humility, may be an unconscious cloak for strong aggressive drives and may even be used as a weapon, as in many cases of maternal overprotection. Similarly, a complicated feeling such as jealousy may have varied components: suspicion of the partner may be a projection of the person's own desires, and what he regards as his own intense love may be rather an intense need to be loved in order to maintain self-esteem. In neurotic illnesses more elaborate disguises of real aims and wishes are sometime apparent.

Such judgments concerning goals, character-traits, feel-

---

1. See, for example, Sigmund Freud, *The Ego and the Id,* translated by Joan Riviere (London, Hogarth Press and the Institute of Psychoanalysis, 1927). For a recent systematic account, see Anna Freud, *The Ego and the Mechanisms of Defense,* translated by Cecil Barnes (New York, International Universities Press, 1946). For a survey covering non-Freudian as well as Freudian schools, see *Psychoanalysis: Evolution and Development* by Clara Thompson, with collaboration of Patrick Mullahy (New York, Hermitage House, 1950).

ings, refer to particular individuals in particular contexts. They have to be established on the basis of careful investigation of the individual's psychological history, and cannot be generalized for the given culture or for all men unless there is further evidence that we are dealing with a culturally invariant or human invariant property. Thus radicalism in a given individual may represent a revolt against the father, but to generalize for all radicalism in this way is to assume that the realistic need for social change in many parts of the world is not itself a sufficient basis for generating radicalism. Again, it may be true, for example, that a Don Juan type in our society will be found invariably to be suffering from intense anxiety, aiming to overcome by a continuous series of sexual conquests, a feeling of inferiority. But it does not follow that this analysis holds for premarital promiscuity in Samoa.[2] Universal judgments are not precluded a priori: it may very well be that intense jealousy always proceeds from a psychological constellation involving a need to be loved in order to maintain self-esteem rather than from a high degree of love. But such judgments have to be empirically established.

The fact that each particular case needs to be separately investigated and established must not obscure the overall generalization that has been established—that such blinds and disguises do occur. It follows that we cannot accept an individual's statement of his ultimate values as absolute ground for conceding them to him as an ultimate indisputable ethical credo.

UNDERLYING LOGICAL ISSUE. There is one logical objection that is sometimes raised to the whole psychological procedure of challenging a man's conscious values in terms of his real values; because of its prevalence in the past, it should be faced at this point. It is the argument that there can be no other "real" values apart from the values that appear in consciousness. At most, it is said, the psycholo-

---

2. Cf. Margaret Mead, *Coming of Age in Samoa* (New York, William Morrow, 1928).

gist is discovering *causes* of the conscious values, or conditions on the basis of which he may predict that the value
in consciousness will disappear under certain altered conditions. If a process of repression took place, does it mean
anything more than that, for example, the wish for parental
warmth of feeling prevailed over the wish that was thrust
aside? When a man in psychological distress comes to a
psychologist for assistance, should we say anything more
than that he is asking help in getting rid of the conflicts
which generate his suffering? That is asking that his values
be in part *changed,* not that any "real" values be discovered
for him or within him. Surely any attempt to tell a man that
he is "really" valuing something else than what he values
is a misuse of language, or else an elaborate attempt to persuade him to adopt the proposed values!

No doubt there is considerable logical work yet to be
done in providing precise analyses for the kinds of psychological judgments we have been considering. But there is
nothing esoteric about the use of such terms as "real values"
and "the realities of the situation." They are continuous with
ordinary usage, and should not be taken to indicate some
superior metaphysical brand of being. Sometimes the reference is simply to a person's mistaken analysis of his own
situation; it may, for example, be clear to the observer but
not to the individual that the latter is desperately trying to
hold something as a goal which he does not want. That we
are dealing with current motivation and not simply past
causal conditions can often be seen from the way the alleged
value is held in consciousness; there is an anxiety-ridden
quality surrounding it, or a compulsive feeling. For example,
where excessive cleanliness appears as a reaction-formation,
the psychologist may say that the person is really fighting
internal tendencies to let go and be dirty, rather than at-
tacking the dirt as a disvalue. This is not refuted by the
fact that the patient lacks insight into his own current motivation. He is in a position comparable to that of a man to
whom a goal has been suggested during hypnosis and who
later pursues it without knowing why. When he has come to

know himself better he may himself see the excessive character of his passion for cleanliness as tied up with an urge to behave differently, as not just an aftermath but an actual continuation of an old old fight with his mother. Another indication of its unreal character is its unrealistic character —the scrubbing may be quite excessive, out of proportion to that required to remove actual dirt. The notion of "real" thus conveys a sense of adequacy or appropriateness to the situation.

Often again, "real" refers to the literal as against the symbolic. Thus the child's fear of animals may be taken to represent fear of his own destructive impulses. This, like the interpretation of dreams, or the theory of art, opens up a wide area of fruitful logical analysis of the senses in which the present content of consciousness may be said to "reflect" phases of what lies beyond its immediate awareness. This may be a large task, and the senses of "real" indicated in this brief summary already present a wide variety. But such inquiries are part of the regular business of philosophical analysis. It is perfectly capable of formulating a definition of values which interrelates qualities in consciousness and dynamic processes in the human being, once it is clear that such a definition is more fruitful than one limited to consciousness alone.

The question concerning the nature of the psychoanalytic process—that is, whether it is essentially discovering the patient's real values or changing him by a play of causal forces—is not, of course, resolved by ascribing a meaning to real values and by seeing the mechanisms of defense which may yield distortion in consciousness. It is possible that in some cases analysis may be removing road-blocks, so that the patient's real values can move ahead. It is also possible, however, that in other cases the patient may have to change his values to resolve his distress. And it is possible, of course, that the one description fits some types of pathology, the other quite different types.

§ The Role of Fundamental Needs

THE CONCEPT OF FUNDAMENTAL NEEDS. In contemporary psychology, the search beyond the individual's conscious preferences has led to the recognition of fundamental needs with which every man in fact has to reckon, whatever the extent of his consciousness of the process. Actually, the concept of fundamental or basic needs is one of a family of concepts, ranging from instincts and drives through dependable motives to acquired value-configurations. Instinct is now reduced to small proportions in the theory of man, and drives are left with a bare residue in some theoretical approaches. But the concept of fundamental needs continues to have a wide and vigorous use.

The analysis of the concept is far from complete in contemporary theory. Logical problems have been raised as to whether we have a right to go beyond actual phenomena of behavior and experiential quality and fashion concepts of the need family that imply potentialities and hidden or as yet unknown mechanisms. This is part of the general problem of maintaining verifiability for propositions that go beyond what is directly experienced. It is now recognized that while such problems impose logical tasks in clarifying need concepts they may no longer be regarded as a bar to their use.[3] Again, it is by now obvious that some normative conceptions are contained in the idea of needs. Even to speak in biological terms of tissue needs—such as hunger, thirst, maintenance of temperature—is to assume survival of the organism as a good. The idea of fundamental psychological needs is correspondingly more complex. It implies some kind of force or operative organization in human beings—the dynamic element traditionally embodied in the drive concept. It assumes the relative invariance of this organization

---

3. For a brief review of such problems as whether needs are to be regarded as "intervening variables" or "hypothetical constructs," see Herbert Feigl, "Principles and Problems of Theory Construction in Psychology," in Wayne Dennis and others, *Current Trends in Psychological Theory* (Pittsburgh, University of Pittsburgh Press, 1951), pp. 179-213.

in men. It sees an unavoidable production of stresses or tensions. And the fundamental character of the need lies in the fact that its continual non-satisfaction, if it does not actually menace survival, will lead to break-down or a major disruption of "adequate" functioning. The implied criteria of inadequate functioning will include large-scale frustration,[4] presence of a permeating anxiety, and so forth.

The fuller analysis of the concept of fundamental psychological needs can come in the long run only with the fuller study of their properties, mechanisms, and related psychological consequences. But a concept that will require continuous refinement in the light of growing knowledge need not in its rough form be a confused concept or a question-begging one. It must, however, be used with constant attention to empirical reference and with logical caution that distinguishes purposes of inquiry: for example, setting forth descriptive invariants in human behavior is quite different from looking for underlying causal mechanisms or providing bases for assuming constant valuations in human life.

SOME PROPERTIES OF FUNDAMENTAL NEEDS. There has been considerable controversy about the properties of fundamental needs, especially about the extent of their fixity, atomicity and "autonomy" with respect to the individual's past development.

On the question of fixity, one extreme position comes from a biological instinct theory that sees original permanent elements in the human organism, readily classified as a list of drives and, in the importance of their satisfaction, as needs. The other extreme is a behavioristic sociology operating with an almost purely plastic conception of the organism. In between are conceptions that see a determinate form arising from the cooperation of organism and fixed features of the environment, or organism and invariant sociopsychological problem-situations.[5]

---

4. See above, pp. 144 ff.

5. For a recent study of opposing views centered on the problem of psychological invariance and change, see Solomon E. Asch, *Social Psychology* (New York, Prentice-Hall, 1952), Chapter 3.

Overlapping somewhat is the issue of atomicity in instincts, drives and needs. Traditional lists gave the appearance of self-sufficient items functioning each on its own, with some mutual conflict or reenforcement. As against this there have been posed organismic and field approaches. Goldstein, for example, regards the appearance of isolated drives as a mark of an abnormal state and reanalyzes drive phenomena in terms of the whole organism seeking self-actualization.[6]

The issue of the degree of dependence on the past to be found in functioning needs has been especially productive of theoretical conflicts. Some views regard all conduct as motivated primarily by the attempt to satisfy certain inflexible drives, inborn or early structured. Allport's principle of the *functional autonomy of motives*[7] is directed against such views. It claims that adult motives become self-sufficient dynamically and are no longer to be regarded, in Freudian fashion, as directed to infantile aims. This issue is much more complex than it seems. In one sense, there is nothing to prevent a psychoanalytic theory from using the degree of autonomy itself as a mark of realistic, as contrasted with neurotic behavior. Or again, if autonomy becomes too complete, so that the "derived need" or "quasi-need" operates without reference to central purposes of the organism, we have, as Goldstein points out, an unnatural isolation.[8] On the whole, these issues have tended to be approached too much in an all-or-none fashion. More careful distinction is needed of the component problems: the persistence of the past in present motivation, the executive role of central purposes, the cultural determinants in need specification, and so on.

The issues of degree of fixity, atomicity and dependence

---

6. Kurt Goldstein, *The Organism* (New York, etc., American Book Co., 1939), esp. pp. 183-207. It is important to note that he allows the concept of needs: "The organism has definite potentialities, and because it has them it has the need to actualize or realize them. The fulfillment of these needs represents the self-actualization of the organism" (p. 204).

7. Gordon W. Allport, *Personality* (New York, Holt, 1937), Ch. VII.

8. Goldstein, *op. cit.*, p. 206.

on the past are all questions capable in principle of empirical resolution. Because they involve broad conceptions of the nature and functioning of the organism which are likely to guide more specialized research, their analytic differentiation is particularly important. Otherwise they tend to get lost as empirical questions, and opposing hypotheses become hardened into methodological counters in a battle of the schools.

RELATION OF NEED AND VALUE.   Given the complex structure of the need concept, there is perhaps no simple relation between need and value. The investigation of the way in which fundamental needs function in human life shows that they provide raw material for many values, often serve as structural components of values, and enter into numerous value-configurations.

Need and value are, however, sometimes contrasted in general on the ground that need satisfaction involves a negative conception of the good as relief from tension, whereas value connotes a positive notion. Dorothy Lee, for example, follows this path in rejecting the concept of basic needs as ultimate in favor of culturally developed systems expressing basic values.[9] She sees a deep penetration of cultural influences in creating needs which we then read back as basic and incorrectly assume as invariant. Such a rejection, however, tends to obscure the variety of possible relationships between needs and values. Whether there are ultimate needs which are sources of major tensions invariable in human life is one question. Whether, if there are such, they are part of the cause in all behavior is a second. Whether this release of tension stands as a goal for all behavior or only for some, is a third question. Whether the negative character of tension release is to be regarded as a prototype of all value, or only of some values, is a fourth. Whether, if this negative character is found in only some values, the existence of certain tensions may still be a necessary condition for the presence of positive values is a distinct further problem. No

9. Dorothy D. Lee, "Are Basic Needs Ultimate?" *Journal of Abnormal and Social Psychology*, 43 (1948), pp. 91-95.

general incompatibility of need and value is pointed to by the answers to such questions.

WHAT ARE THE SPECIFIC BASIC NEEDS? It is generally agreed by this time that human needs are much more universal than the diverse forms of socially conditioned desires would seem to suggest. This lesson is common to most contemporary psychological schools. Modern Behaviorism in its early days under John B. Watson had given a powerful impetus to ethical relativity in the view that initial tendencies were limited to a few simple reactions and that almost anything could be made of any normal child. But contemporary Behaviorism, as represented in Clark Hull and E. C. Tolman, finds no difficulty in using concepts of primary needs or mapping innate purposive drives.[10] And, of course, outlooks that draw upon psychoanalytic theory or in general upon clinical work constantly operate with a framework cast in terms of needs. Even those who prefer to think in terms of basic value systems rather than basic needs have to work out comparable concepts to play a similar role; for example, they will recognize that if certain values are neglected, there are definite types of disruption that arise in the lives of the persons concerned.

The central psychological problems concerning particular basic needs arise when we go beyond the merely biological ones of food and sex, although the interpretation of the latter is intimately involved in these issues. Let us take the contrast of aggressive power needs and affiliative interpersonal needs as a central problem from which fundamental ethical consequences would be derived.

Some aspects of the problem of a basic drive or will to power have been dealt with in outlining the various threads in the pattern of ethical relativity and in the discussion of the theory from which Freud derived an unavoidable aggression in man; the power motif in social and historical

---

10. See, for example, the treatment of needs in Edward C. Tolman, "A Psychological Model" in *Toward a General Theory of Action*, edited by Talcott Parsons and Edward A. Shils (Cambridge, Harvard University Press, 1951), pp. 277-361.

theory will be considered below.[11] The will to power as the basic phenomenon in man was propounded by Nietzsche. In psychoanalytic theory Alfred Adler gave it the same central role that Freud had assigned to sexuality.[12] But while there are many factors in the early situation of the child which give prominence to power feelings and power relations, and while "primary sadism" is also now widely given a role in the early developmental picture, exaggerated power-strivings in the adult have become more and more suspect as pathological phenomena in psychological theory. The occurrence of the power drive as central and persistent in the consciousness and behavior of an individual would be seen as dominantly an insecurity phenomenon.

In social interactions, the struggle for what we rather abstractly call "power" is usually an attempt to secure specific goods or resources, to exploit other people for concrete gains. These phenomena involve serious moral problems, but it has little to do with the psychological power drive alleged to constitute the basis of individual striving. Alexander sighing for more kingdoms to conquer, or Cecil Rhodes looking at the moon and wishing he could annex it are definitely aberrant phenomena. Similarly, the experiencing of all interpersonal relations in terms of aggression or submission, the imparting of sado-masochistic quality to all affectionate relationships, are definitely pathologic phenomena. That politics should become widely interpreted by such relationships is itself a *social* datum requiring social and historical explanation.

By contrast, the existence of a need for warmth, affection, of affiliative (belonging) needs, has been increasingly recognized, in spite of the controversy whether it is to be regarded as strictly primary or a secondary outgrowth. G. W. Allport states it in its widest moral implications when he

---

11. Above p. 24 and pp. 127-130; below pp. 250-254

12. Adler stresses the inferiority feelings inherent in the human situation. Cf. his *Understanding Human Nature*, translated by W. B. Wolfe (Garden City, N. Y., Garden City Publishing Co., 1927), Book I, Ch. 5. For a clear statement of Adler's position, see Patrick Mullahy, *Oedipus, Myth and Complex* (New York, Hermitage Press, 1948), Ch. 5.

says: "It seems probable that every child in every nation, the world over, at a time when he is most plastic, wants security, affection, and an affiliative and comprehending relation to the surrounding world. It is conceivable that the same basic intentions exist in most adults, although thwarting and perversion of this relationship have engendered a vast amount of hatred, emotional instability, and warlike impulse."[13] Behind such judgments lies the accumulated knowledge of the way in which frustration breeds aggression,[14] the role of scapegoating in social as well as individual life, clinical experience with sadism, the emotional effects of repressive adult behavior to children, and a whole vast array of experience and established theory. In short, the psychological analysis does not merely list the need but shows the manifold ways in which it finds devious expression where thwarted.

IMPACT OF THEORY OF NEEDS ON ETHICAL RELATIVITY. The merely general assertion that there are fundamental human needs narrows down indeterminacy only by contrast with the view that man is almost wholly plastic. The discovery of particular needs such as affiliative tendencies and their pervasive effect when distorted does, however, carry us forward to a remarkable degree. It provides a basis for ethical judgment between many alternatives hitherto regarded as equally arbitrary or a matter of cultural "taste." A good illustration is to be seen in the revolutionary effect

---

13. "Scientific Models and Human Morals" (Divisional Presidential address before the first meeting of the Division of Personality and Social Psychology of the American Psychological Association, Sept. 4, 1946), in *Psychological Review* 54, 4 (July 1947), 188. For a basic comparative study of the effects on children of lack of maternal warmth in institutional conditions, see René A. Spitz, "Hospitalism. An Inquiry into the Genesis of Psychiatric Conditions in Early Childhood" and "Hospitalism: A Follow-up Report" in *The Psychoanalytic Study of the Child* (New York, International Universities Press), Vol. I (1945), pp. 53-74, Vol. II (1947), pp. 113-117. See also J. Bowlby, *Maternal Care and Mental Health* (World Health Organization Monograph No. 2, 1951). A brief treatment of the question will be found in Ashley Montagu, *On Being Human* (New York, Schuman 1950), pp. 55-71.

14. See above, pp. 145 ff.

of psychological knowledge on what is regarded as right or wrong in child care. How hollow today is the argument in favor of stern discipline in the cradle in order to produce a disciplined race. Lecomte du Noüy, in his *Human Destiny,* regards a mother as sinning against her child if she does not let him yell awhile before offering him the bottle; for his future disciplined character—and therefore the human race —is endangered thereby![15] Now before the growth of the psychological knowledge it all depended. If you wanted a Spartan society you advised Spartan discipline from the first moment. You got Spartan courage—although in modern terms you are more likely to get Prussian militarism in which obedience to those above is coupled with compensating aggressiveness to those below. You can still advocate such an upbringing if you are planning a life of warfare. But this you can no longer do: you cannot go ahead blithely on the assumption that you are meeting the needs of the persons concerned rather than thwarting them and hammering them into a distorted pattern, even if you start from the very first moment so that the persons have never consciously wanted anything else.

Now such consequences in diminishing indeterminacy are sometimes thrust aside by the claim that affiliative tendencies are secondary and derivative rather than primary needs. All needs, it is said, are inherently egoistic. They are demands of the individual for release of tension, and the relation to the human environment, just as to the material surroundings, is fundamentally exploitative. And since tension is a phenomenon localized in the individual and its release is generally productive of pleasure, this prominent personal reference in the goal is not unlike that in traditional hedonism. The outcome is to sweep the whole of a "needs" approach into the camp of an individual arbitrary relativism on the basis of an unavoidable egoism.

If such an approach is cast in empirical terms, then it

---

15. Lecomte du Noüy, *Human Destiny* (New York, Longmans, Green, 1947), pp. 209 ff. Compare with this, L. K. Frank, "The Fundamental Needs of the Child," *Mental Hygiene* XXII (1938), pp. 353-379.

can only be equivalent to some of the hypotheses about need properties and specific needs that were considered above, such as their lack of any autonomy and inevitable infantile reference, their basis in a power-seeking internal mechanism, and so forth. These can be met in essentially empirical terms by the kind of evidence that exhibits the genuinely inter-personal and social quality of the affiliative needs, and pro-vides an accurate account of the processes that go astray to produce an exploitative or grasping quality of feeling. On the other hand, if the approach to needs as inherently ego-istic takes some generalized theoretical form, it calls for a careful analysis of the concept of egoism itself. And to this in turn contemporary psychology has itself contributed.

§ Psychological Evaluation of Egoism

PSYCHOLOGICAL CLARIFICATION OF EGOISM.   The egoistic view of man's nature is not refuted directly by psychological data. It is rather cast down from its a priori formulations and told to turn itself into a respectable hypothesis capable of being tested, or else into one pattern of possible human relations capable of being evaluated among alternatives. This is achieved by showing that particular psychological assumptions have always been inherent in the egoistic ap-proach and by exposing and criticizing those assumptions in the light of contemporary knowledge.

The psychological basis of egoism as a pillar of an arbi-trary ethical relativity has always been an atomic concep-tion of the self. Why should *I*, it is said, pursue any aim unless it is a benefit to *me?* But what is a benefit to me de-pends in the first place on what I am found to be. If I am a pleasure-seeking animal, then Bentham can present me with a ruler and calculus for measuring and anticipating pleasure. If I am a God-fearing animal, then a calculus of anguish and terror might be more serviceable. If what I seek is self-realization, then T. H. Green will instruct me how to look within myself, so that by the time I am through my self will be coextensive with the social institutions of my age.[16] Or if I am just I, then Max Stirner will help me detach foreign religious or liberalistic ideals, and leave me with what is properly my own.[17] But whether I am then a lord or completely impoverished again depends on what I really am.

The lack of an empirical conception of the self in traditional egoism has allowed extreme vagueness both in its defense and in attack upon it. The ethical discussions of the 19th and 20th century on this topic are rarely an improvement over those of the 17th and 18th. For example, when John Stuart Mill, in his *Utilitarianism,* sets out finally to prove that men desire nothing but "that which is a pleasure to them or of which the absence is a pain" he poses the inquiry as if it were a matter of fact to be settled by empirical correlation of phenomena, and then ends by discovering that to desire and to find pleasant are just two different names for the same psychological fact.[18] Or when Moritz Schlick wishes to refute the formulation that a man pursues his own pleasure, he argues that there is no such object of motivation as the idea of a pleasant personal state; I can only *feel* pleasure and pain, not have them as objects of thought or imagination![19]

The special virtue of contemporary psychology on this question is that it operates with a verifiable account of the self and the ego. It makes careful experiments distinguishing the person who pursues a goal or completes a task with an interest that is task-oriented from one whose interest is in the feeling of self-enhancement that comes from doing the task.[20] It does not equate the ego indiscriminately with all

---

16. T. H. Green, *Prolegomena to Ethics* (Oxford, Clarendon Press, 1883).

17. Max Stirner, *The Ego and His Own,* translated by Steven T. Byington (New York, Boni and Liveright).

18. J. S. Mill, *Utilitarianism,* Ch. 4.

19. Moritz Schlick, *Problems of Ethics,* translated by David Rynin (New York, Prentice-Hall, 1939), p. 71.

20. See, for example, "An Experimental Study of the Role of the Ego in Work. I. The Role of the Ego in Cooperative Work" by Helen Block Lewis, and "II. The Significance of Task-Orientation in Work" by Helen

that an individual does or says or feels. Instead, it identifies
the ego with a definite part or role in the self, as in the
Freudian distinction of id, ego and superego. Where the
pleasure principle is employed in the theory of motivation,
it is not the cover-all which assumes that pleasure is the sole
object of striving. Instead it is used in some restricted way
as a definite pattern, like the tendency of tension to seek
immediate discharge. In this sense it is contrasted with
deferment in the light of environmental demands, which is
a separate principle (the reality principle). An ego and a
pleasure principle that are empirically defined make possible
a fruitful study of egoism as a specific phenomenon.[21]

EVALUATION OF EGOISM. Egoism can be defined in verifiable terms in one of several ways. It may be some type of
distinguishing attitude in interpersonal relations. It may be
a kind of configurational quality of character. Or it may
denote a set of social relations supporting a specific type of
character. Schlick, in his concrete analysis of egoism, heads
in a scientific direction by defining egoism as inconsiderate-
ness. The selfish man is sometimes described on a wider
range in psychological writings today as the *exploitative*
character—one who takes a "what's in it for me?" approach
in every conceivable situation, who sees and feels the other

Block Lewis and Muriel Franklin, *Journal of Experimental Psychology*,
34, 2 (April 1944), 113-126, and 34, 3 (June 1944) 195-215.
    21. Concern with empirical study of the ego and kindred concepts,
such as the self has grown during the past decade. See, for example: Gordon
W. Allport, "The Ego in Contemporary Psychology," *The Psychological
Review* 50, 451-478 (Sept. 1943); Muzafer Sherif and Hadley Cantril,
The Psychology of Ego-Involvements (New York, J. Wiley and Sons,
1947); Anna Freud, *The Ego and the Mechanisms of Defense,* translated
by Cecil Baines (New York, International Universities Press, 1946); Heinz
Hartmann, "Ego Psychology and the Problem of Adaptation" in *Organiza-
tion and Pathology of Thought,* selected sources, translation and com-
mentary by David Rapaport (New York, Columbia University Press, 1951);
Solomon E. Asch, *Social Psychology* (New York, Prentice-Hall, 1952), Ch.
10. Anthropological resources are brought to the study of the self in: A. I.
Hallowell, "The Self and Its Behavioral Environment," in *Explorations,
Studies in Culture and Communications,* II (April 1954); Dorothy Lee,
"Notes on the Conception of the Self Among the Wintu Indians," *The
Journal of Abnormal and Social Psychology* 45, No. 3 (July 1950).

person only as a source for deriving some material or psychological gain. Such an identification cannot here be explored in detail. But its psychological evaluation in the extreme case seems clear enough. A hardened egoism in which there was in fact (not just in theory or argument) no feeling relation other than an exploitative one to other people would appear to be the product of a distorted development. Some psychological descriptions regard it as a fixation at an earlier (anal-sadistic) developmental point; others as a narcissistic withdrawal of love for others and turning it on the person. Some point to an anxious insecurity as its base. Erich Fromm, in his *Man For Himself,* goes so far as to argue that a man who is incapable of loving others is really incapable of loving himself, that his apparent self-love is simply a driving anxiety. Whatever be the correct psychological analysis of the really hardened egoist, the overwhelming tendency is to regard such an adult as a pathological specimen rather than as the natural man unseduced by idealistic illusions and social blandishments.

It does not follow, of course, that the type of egoism normal to a culture in which the realistic path to success is outdoing others has the same pathological character as the egoism of the completely hardened egoist. (How extensive the pathological element is in our society cannot be told readily, especially since it will be covered by socially acceptable practices.) An extreme competitive-success society establishes exploitative conduct as a realistic mode of achieving desirable goals. It penalizes thereby those whose attittudes in interpersonal relations have a quality of warmth and concern for others—the kinds of qualities contemporary psychology has found desirable in sound human personal development. Thus a man who is moved by humane considerations in a business deal, to his financial detriment, is often regarded as "weak." To have revealed such consequences of egoistic social patternings is one of the major social contributions of psychology today.

A full-length exposition and critique of egoism would itself require a whole book. But in any case the psychologi-

cal insistence on precise formulation has made a vague egoism that would cover any form of behavior out of date. And
the psychological denial of naturalness to egoistic character
has removed its assumption that it is unavoidable. The net
result is that egoism can no longer act as a base for ethical
relativity. The partisan of egoistic patterns in individual
and social life today has a hard enough job to justify his
particular selection in comparison with other possibilities;
for it is no longer self-justifying. He is much too busy to act
as a complacent pillar for an individualistic ethical relativity.

§ Impact of a Theory
of Personality Development

THE CONCEPT OF MATURITY. A limitation of indeterminacy similar to that following from a recognition of fundamental needs is secured by providing a detailed account of
personality development. There is as yet considerable controversy in this area. As the scientific task is progressively
accomplished it will become increasingly possible to formulate an ethical concept of *maturity*.

Such a concept is to be distinguished from the more
generic idea of growth, which is a kind of synthesis of
increase and novelty. The belief that there is a growth-
requirement in the human make-up takes different forms
and some of them seem to embody specific values. For example, there is the view that a normal part of man is the
urge to have ever new experience and not to be satisfied
with simply repetition of old ways.[22] Again, a growth-re-
quirement may be seen as a specific desire for improvement,
the unwillingness to have things remain as they are, stemming from a general goal of the enhancement of value in
experience.[23] In such conceptions one would have to ask

22. Cf. H. A. Overstreet, "The Growth Imperative," in *Freedom and
Reason,* edited by Baron, Nagel and Pinson (Glencoe, Illinois, The Free
Press, 1951)
23. Cf. Hadley Cantril, *The "Why" of Man's Experience* (New York,
Macmillan, 1950).

how far we are dealing with psychological invariants, how far with special value forms of a modern open world, reflecting historical and cultural elements in the idea of progress. That specific concepts of growth require evaluation is clear enough from the familiar fact that there may be harmful growths as in cancers, or excessive growth.

The concept of maturity, on the other hand, aims to articulate the norms inherent in a scientific account of personality development. If there is no such scientific picture, no determinate structure in that process, then the concept of maturity reflects arbitrary valuations. But if there is a definite structure it does offer a base for diminishing indeterminacy in evaluating character-traits, establishing virtues, and even perhaps to some extent assessing social patterns.

ITS USE IN EVALUATION.   Let us take as an illustration, the question of male homosexuality, discussed by Kinsey. He seems to regard it as a statistical question, so that when he finds some 37% of American males have had at least one homosexual experience (without even inquiring where it represents chance experience or experiment or lack of other possible outlet, and where stable pattern of preference), he endows it with the status of near-normality.[24] Similarly, the cultural relativist has pointed to the fact that some societies standardized homosexual patterns. The ethical implication seems thus to be that it is just a question of cultural selection, unless the objection to homosexuality is biologically cast in terms of reproduction of the species. How does a depth psychology with a theory of personality development here aid moral evaluation? It does not stop with the behavioral phenomenon, but seeks some account of its underlying dynamics. Thus Fenichel takes the essence of male homosexuality to be not the surface preference of male for male, but the underlying fear of the female.[25] This is explained in terms of early developmental problems—fixation at an earlier point or exaggeration of an otherwise normal

---

24. A. C. Kinsey, W. B. Pomeroy, and C. E. Martin, *Sexual Behavior in the Human Male* (Philadelphia, W. B. Saunders Co., 1948), p. 623.

25. Fenichel, *op. cit.*, p. 330

phase. This is not offered as merely genetic explanation; it probes to the present quality of the person's behavior and shows its fear-reaction components. We cannot here consider the particular evidence for the hypothesis. But it does show how a purely psychological account can, to the extent that it is confirmed, completely restructure the problem of evaluation. In general, to value maturity as against immaturity will not be an arbitrary preference if it can be shown that in fact the desire not to grow up represents a fear reaction rather than an extreme qualitative appreciation of the earlier stage lasting into the later.

APPLICATION TO VIRTUES AND CHARACTER-CONSTELLATIONS. Such an approach has an extended effect on the whole traditional treatment of the virtues and the moral ideal. It becomes increasingly difficult simply to line up opposites—passive and active types, egoistic and affiliative, authoritarian and democratic, competitive and cooperative —and treat them as styles of life and character, with decision among them a matter of individual or cultural taste. Take, for example, a recent psychological study of the anti-democratic personality, and consider what ethical consequences follow on the assumption that it is established.[26]

The anti-democratic personality was identified initially in ideological terms, the ideology studied being anti-Semitism. The trends found here—"its generality, stereotyped imagery, destructive irrationality, sense of threat, concern with power and immorality, and so on"[27] were found to hold in general for ethnocentrism, which involves both rejecting the out-group and a submissive idealization of the in-group. This in turn, studied in terms of personality, was found correlated with such variables as clinging to conventional

---

26. T. W. Adorno, E. Frenkel-Brunswik, D. J. Levinson, R. N. Sanford, *The Authoritarian Personality* (New York, Harper, 1950). Summarized in "The Anti-Democratic Personality" in Newcomb, Hartley and Others, *Readings in Social Psychology* (New York, Holt, 1947), pp. 531-541. For a critique of this study, see *Studies in the Scope and Method of "The Authoritarian Personality,"* ed. Richard Christie and Marie Jahoda (Glencoe, Ill., Free Press, 1954.)

27. "The Anti-Democratic Personality," *ibid.*, p. 535.

values, authoritarian submission and aggression, an opposition to an insightful view of people, stereotyped superstition, cynicism, a readiness to imagine strange destructive forces at work in the world, and so forth.[28] Further study looked to the ethnocentric individual's intra-familial relations, to correlations of ethnocentrism with different types of values (e.g., exploitative or affectionate attitude, hierarchical or equalitarian conception of human relations), and the conclusion pointed to was the existence of an anti-democratic personality pattern, embracing both personality and ideology.

A causal hypothesis is suggested in such questions as "Does ethnocentrism help the individual avoid conscious ambivalence toward his family by displacing the hostility onto out-groups (the morally 'alien') and thus leave in consciousness exaggerated professions of love toward family and authority? Do high scorers on the F scale (who are usually also ethnocentric) have an underlying anticonventionalism, in-group- and family-directed hostility, a tendency to do the very things they rigidly and punitively oppose in others?"[29] As a causal hypothesis this would not explain the rise of ethnocentrism or anti-Semitism in psychological terms (as against socio-historical terms) but would explain why certain individuals took or kept hold of it more readily than others, or gave it a more central role in their lives.[30]

To the degree to which such studies provide scientific conclusions, they make possible an ethical evaluation of constellations of traits in personality. The anti-democratic personality would thus cease to be simply one pattern of

---

28. *Ibid.*, p. 536.

29. *Ibid.*, p. 537.

30. A good brief summary of the social history of the authoritarian type of ideal is given in Fenichel, *op. cit.*, p. 587. Fenichel points to its undisputed sway in feudalism, the rise of its opposite in capitalism, the reappearance of authoritarian necessities as subsequent development of capitalism created a majority of people "who had to be kept contented in relative frustration and dependency." As the single individual's activities become hopeless "regressive longings for passive-receptive regulation come to the fore again. Old feudalistic ideas are revived and even increased, and the result is a mixture of ideals, conflicts, and, later, neuroses."

preference in a moral ideal, capable of arbitrary ultimate selection, and instead be viewed as an immature one. The resulting moral judgment is strengthened by further positive investigation into democratic attitudes and the conditions that support them, as in the now classic experiments of Lewin, Lippitt and White on democratic and authoritarian structuring of children's play groups.[31]

§ Growth of Psychological Criteria

for Normative Concepts

NORMALITY AND NATURAL AS AGAINST UNNATURAL BEHAVIOR. The combined knowledge of mechanisms of defense, fundamental needs and processes of personality development opens the way to more significant use of certain concepts which have often seemed in the past to be chiefly vehicles for arbitrary cultural valuation.

Notable among these are the concepts of what is normal and what is natural. These have a long philosophical history in which their varying use reflects underlying theories of man and his relations to the world. This at least we may conclude today, that such concepts are no longer to be equated simply with a preponderant statistical frequency or a pure cultural bias.[32] They presuppose both a theory of needs and a theory of development. A full analysis of "human nature" would show in terms of a theory of need mechanisms, how what is statistically most frequent expresses human nature under favorable conditions but diverges from it when men are driven to compromise with what they need under conditions of hardship or conditions not conducive to mature development. It would show how

---

31. Summarized in Newcomb and Hartley, *op. cit.* pp. 315-330; for more detailed bibliography, *ibid.*, p. 315.

32. Cf. Henry J. Wegrocki, "A Critique of Cultural and Statistical Concepts of Abnormality," *Journal of Abnormal and Social Psychology*, 34 (1939), pp. 166-178. Reprinted in *Personality in Nature, Society, and Culture*, edited by Kluckhohn and Murrary (New York, Knopf, 1948), pp. 551-561.

need concepts enter into definitions of specific ethical terms, and how both needs and statistical or cultural norms are capable of evaluation.[33]

MENTAL HEALTH. With the advance of psychological knowledge, it is possible that the concept of mental health may be given an empirical meaning comparable to that of the biological ideal of physical health. The concept of physical health has itself broadened and become more positive, shifting steadily from a narrower goal of removing illness to an ideal of securing definite conditions based upon knowledge of the requirements of physical well-being. For example, the judgment that rickets is an evil was well-grounded long before the judgment that regular diet should contain such-and-such vitamin balance. Similarly there is no limit to the determinacy which the growth of psychological knowledge can bring to the conception of mental health as it becomes increasingly positive. In fact, the likelihood is that the concepts of physical and mental health will merge with the fuller integration of psychological conceptions in medical theory.

At present, the negative approach still prevails in psychological criteria of health. Literally, more is known about the conditions of pain than about the conditions of pleasure. Such criteria as breaking-points and inability to function, intense anxiety, difficulty in interpersonal relations, sense of compulsion, are all definite enough. They are not mere absence of pleasure. But as means of formulating a standard of psychological health as a good, they are negative, that is, states to be avoided as evil. This approach can be seen in the actual results of the science in its present stage. For example, Fenichel points out that psychoanalytic knowledge of the "successful repressions" is still tentative: "The conditions creating reactive character traits are much better known than those responsible for the sublimation type. It can only be stated that the absence of conditions favoring

---

33. For evaluation of drives, see above pp. 137 ff. For attempts to establish trans-cultural bases of evaluation in terms of needs, see below pp. 220-225.

the development of reactive traits is the main prerequisite for the building of sublimations."[34]

The fact that the approach is dominantly negative from an ethical point of view should not lead us to underestimate its contributions, especially on the issues of ethical relativity. Not only can specific preference paths be evaluated (as in the examples given above), but generalized criteria can be partially formulated, and superficial moral judgments corrected. For example, concepts such as "impoverishment of personality" can be employed to recognize even in relatively well persons the state in which a great part of one's energy is used for defenses through repression; similarly, one can recognize that every man has his "breaking point" and use as a criterion the ease with which this point is reached.[35] Again, one can by use of psychological data correct the ethical judgment frequently made in the hedonistic school that pleasure is pleasure wherever you find it. Thus Fenichel states of perversion: "Some people think that perverts are enjoying some kind of more intense sexual pleasure than normal people. This is not true. Their discharge has become possible after hindrances and through distortions only and is therefore necessarily incomplete. They are, as Freud states, poor devils who have to pay a high price for their limited pleasure."[36] In general it becomes impossible to follow a simple hedonism. The hedonic criterion can no longer be used as one capable of independent application without reference to the internal economy of the individual.

HAPPINESS.   If we add to these results the attempts which we looked at in the last chapter to distinguish and evaluate types of pleasure,[37] it is clear that contemporary psychology is aiming even beyond a positive conception of mental health, to a positive empirical theory of happiness.

---

34. Fenichel, *op. cit.*, p. 471. Cf. the discussion of sublimation, pp. 141-143.

35. Cf. Fenichel, *op. cit.*, pp. 121-122. Cf. also Karen Horney, "The Paucity of Inner Experiences," *The American Journal of Psychoanalysis*, XII, 3-9 (1952).

36. Fenichel, *op. cit.*, p. 328.

37. P. 135.

When we recall the role that happiness has had as a goal in ethical conceptions of the good, it is apparent that its successful empirical study will have momentous consequences. Philosophical analyses have made distinctions between general feelings of contentment or euphoria and the well-being (eudaemonia) which involves a basic fulfillment; but on the whole, concern has been rather with necessary conditions for happiness than with empirical criteria for its identification. But this task should not be too far beyond the reach of contemporary accumulated materials. At least in the case of children the question of the marks of a happy child may properly take its place among the various studies of child training and developmental processes, to be explored by psychology in cooperation with the social sciences.[38]

PROGRESS. Even beyond the theory of happiness lies the possible contribution that psychology may make to a general theory of progress. It could not, of course, provide such a theory wholly by itself. But it could map the general features of advance on the psychological side in the light of its conceptions of maturity and happiness. Flugel, for example, lists a number of aspects which can be used to judge the psychological progress of a society or of civilization.[39] They are the progress from egocentricity to sociality, from unconscious to conscious, from autism (fantasy) to realism, from moral inhibition to spontaneous goodness, from aggression to tolerance and love, from fear to security, from heteronomy to autonomy (from other-rule to self-rule). Although many an ambiguity may lurk in such concepts, and many a social value be incorporated in a psychological no-

---

38. Some elements in such a concept geared to the developmental process are implicit in Erik H. Erikson's account of eight stages in the growth of man, in his *Childhood and Society* (New York, Norton, 1950), Ch. VII; especially in comments on such questions as smiling in infancy, feelings of trust and confidence, possibilities of love and intimacy, exuberant enjoyment of new locomotor and mental powers, pleasurable accomplishment, etc.

39. J. C. Flugel, *Man, Morals and Society* (New York, International Universities Press, 1945), Ch. XVI.

tion (which would then require more than psychological justification), the list does indicate the major path of expectation. For these are not just words; they are increasingly given empirical or operative meaning in the actual growth of psychological knowledge.

§ Psychological Evaluation of "Amorality"

FORMULATION OF THE PROBLEM. Most of the psychological materials so far dealt with impinged on the ethical problem of diminishing indeterminacy and lessening arbitrary preference in the following ways: by showing lack of insight in specific judgments, by establishing a set of fundamental needs whose satisfaction was a good, by showing the neurotic character of the power drive and the pathological nature of hardened egoism, by exhibiting a developmental pattern so that any regression or fixation was in some sense denying one's own "growth-imperative," by trying to find empirical marks of such concepts as happiness. This is an imposing array, but it is time now to face any issues of ultimate ethical indeterminacy left in these analyses themselves. Let the objector whom we keep in reserve as the guardian of the critical spirit pose the following complex illustration. Suppose a man who has insight into what he is doing, that is, he does not to a marked (abnormal) degree employ escape mechanisms. Suppose that he denies the fundamental need of affectionate relationships. Suppose on discussion he accepts the developmental pattern but denies that for him it constitutes a growth imperative. "You are showing me," he will add, "possibilities that I might have actualized but did not. Other causes intervened in my case. Is this different from saying that I once had the capacity to be a musician or a mathematician, but failed to develop them? Must I forever be regarded as an immature musician rather than as not a musician at all? Surely the test is whether you still find any such aspirations within me, conscious or unconscious." Suppose further, then, he claims to be happy. Suppose again that he has no neurotic power drive, but is ready to seize power realistically for whatever he happens to desire. And finally, suppose he is an egoist in the sense of disregarding anyone else's claims in contrast with his own desires, but not in the sense of an anxiety to outdo and exploit others. Moreover, his desires are never other-oriented, in terms of affectionate feeling. Can psychology provide evidence that his pattern of life is evil or less good than that of the more traditional good man? Will such a man be dismissed as simply "morally blind"? Or will he be recognized not as *wrong* but as merely different, *ultimately different?*

THE SEARCH FOR SPECIMENS.   In such a case, the scientific spirit shows itself by affirming neither of the possibilities, but instead by attemping to locate, describe and explore the phenomena to which attention has been called. Are there such persons? What is the probability of the conjunction of such characteristics? In short, are they real, or merely ghosts from the cupboard of the stubborn man? Is there such a phenomenon as moral blindness, or is it simply part of the language of invective of the absolutist? After such questions are investigated it will be time to consider what follows from the mere conception of such a being.

That there are individuals who seem to fit the bill in at least some of the respects indicated is pretty clear. Their classic analysis is that of the tyrant in Plato's *Republic*. In fact Plato is consciously tackling the psychological problem at issue. He defines the tyrant as the man who is guided by impulse, whose conception of the good is simply the satisfaction of his immediate impulses (which do not reckon others), and who amasses the power and means to satisfy his demands. Plato, having dissected the soul into reason, spirit or will, and appetite, describes the tyrant as one ruled dominantly by the third part. And finally, he attempts to establish that the tyrant is the unhappiest of mortals. Plato's tangling of metaphysical and value considerations with his psychological account probably explains why his problem has been comparatively neglected. Another reason is no doubt the subjectivization of pleasure noted above, which led to the ready assumption that the tyrant had his own tyrannical happiness.

MORAL BLINDNESS. The conception of moral blindness has had a comparably mixed history. Plato's tyrant is morally blind in the empirical sense that he seems to have no guilt feelings. Aristotle distinguishes carefully between the brutish man, whom he regards almost as a literal beast, and the vicious man who follows laws but ones that are different and immoral. Epictetus speaks of a man becoming hardened like a stone, his sense of shame and self-respect deadened, his moral sense petrified. Kant talks of corruption or depravity in the human heart, "the propensity of the will to maxims which prefer other (not moral) springs, to that which arises from the moral law."[40] It is chiefly intuitionist philosophers who use the metaphor of moral blindness, complete lack of sensitivity or moral insight.

THE PSYCHOPATHIC PERSONALITY. Contemporary psychiatry has come increasingly to identify the type of individual we are dealing with as the *psychopathic personality*. The empirical facts about him are that he has no apparent guilt feeling, and therefore seems to lack altogether what may be called a moral sense. He has accordingly none of the ordinary scruples in violating the customary moral code. Case histories show over and over again the complete readiness to lie, to take what he wants, to be ruthless towards others, not necessarily in any sadistic sense, but more in the sense that it does not matter, to be unable to resist impulse. Sometimes, if circumstances are propitious, such a person may lead a well-adjusted life, in the sense of maintaining stability and succeeding socially, but more often not.

There is considerable difference in the classification of the psychopathic personality. Older conceptions of a well-defined type of "moral insanity" rested on assumption, found

40. In his essay "On the Radical Evil in Human Nature," Abbott's *Kant's Theory of Ethics* (London, Longmans, Green, 1879), p. 397.

among moral philosophers, of a separate moral sense that could be diseasesd.[41] Contemporary classifications are as varied as psychosis (but differing from all types now recognized), impulse neurosis, and a separate condition falling outside the more well-defined groupings.[42] Some take the essence of psychopathy to be prolonging infantile patterns into physiological adulthood.[43]

Despite these variant diagnoses, considerable knowledge has been gathered on the etiology of the psychopathic personality. A major determinant seems to be an extreme absence of satisfaction of the need for tenderness and love in childhood, and a consequent lack of the kind of identifications in which a superego is formed. Many stress the difficulties of successful therapy, but some see the possibility of emerging from the state under favorable conditions in which superego surrogates may develop.

THE PROBLEM OF CONSCIENCE IN THE PSYCHOPATHIC PERSONALITY. Most relevant to our present inquiry is the analysis of the apparent lack of guilt in the psychopathic personality. On this question there is marked disagreement. Some take the phenomenon at its face value, insisting that conscience is completely absent in pure psychopathy and setting aside cases of disturbed conscience as secondary.[44] Others insist, however, that guilt feeling and anxiety characterize all cases although forms they take may be sometimes dis-

---

41. For a historical survey of the concept of the psychopath, see Sidney Maughs, "A Concept of Psychopathy and Psychopathic Personality: Its Evolution and Historical Development," *Journal of Criminal Psychopathology*, 2 (1940-41), pp. 329-356 and 465-499. For a recent study of different approaches, see Milton Gurvitz, "Developments in the Concept of Psychopathic Personality (1900-1950)," *The British Journal of Delinquency*, Vol. 2, No. 2.

42. See respectively, Hervey Cleckley, *The Mask of Sanity* (St. Louis, The C. V. Mosby Co., 1941), pp. 258-259; Otto Fenichel, *The Psychoanalytic Theory of Neurosis* (New York, Norton, 1945), pp. 373-375; D. K. Henderson, *Psychopathic States* (New York, Norton, 1939), p. 136.

43. R. M. Lindner, *Rebel Without Cause* (New York, Grune and Stratton, 1944), p. 2. Lindner also suggests physiological factors as relevant (pp. 10-11).

44. Cf. Ben Karpman, "Conscience in the Psychopath: Another Version," *American Journal of Orthopsychiatry*, XVIII (1948), p. 458.

persed and have to be tracked down.[45] Some even find the psychopath so burdened with guilt as to be literally seeking punishment.[46]

There is corresponding difference on the related issue of whether the psychopath can be happy. The picture of the happy narcissistic psychopath would seem to rest largely on the assumption that guilt feeling is absent, only external punishment is feared, and impulse is gratified without any consideration for others. The opposing view finds the irresistibility of impulse in the psychopath different from that of normal instinctual drives; it calls attention to lack of insight and emotional response and a general flattening or hollowing in affect.[47]

IMPLICATIONS FOR ETHICAL RELATIVITY. Obviously it is not the function of our present ethical inquiry to pass judgment on these conflicting psychological conceptions. But the accumulation of the evidence does make it increasingly difficult to treat a complete and hardened amorality as a self-sufficient type, simply an ethics on its own of a pure desire-expediency type. We can no longer take such a man at his word when he says that being incapable of feeling guilt he is untroubled, or that having no real love for others there are no unsatisfied interpersonal needs (other than expediency relations) within him, or that he is at peace with his impulses, or that he has insight into his own needs and values. If this scepticism on our part is scientifically justified, then the whole force of the relativist's argument—"Suppose there were a happy Platonic tyrant; how would

---

45. Fenichel suggests several mechanisms that might explain the sidetracking of guilt-feeling: "isolation" of the whole superego so that there is yielding to impulse in one context and displaced "remorse" in another context, reaction-formations to guilt-feelings, "idealization" of instinct activity bringing it in line with superego demands (*op. cit.*, pp. 373-375, 166, 504). Cf. Phyllis Greenacre, "Conscience in the Psychopath," *American Journal of Orthopsychiatry*, XV (1945), p. 506.

46. Lindner, *op. cit.*, p. 8.

47. Fenichel, *op. cit.*, pp. 367, 374; Cleckley, *op. cit.*, p. 258. Cleckley insists that the psychopath is not to be compared to a man who sacrifices ordinary satisfactions for an enthralling sin that ruins him (*ibid.*, p. 70), and denies that even drinking brings him any special satisfaction.

you judge his morality incorrect?"—utterly evaporates. For the person even after the psychopathic state has developed has not, so to speak, cut his moorings to leave only a causal relation but not a functional relation to his past; what we have is a distorted pursuit of the same needs which were never satisfied.

If, on the other hand, this scepticism is not justified, and we have here a stabilized type sufficient unto itself, the scientific evidence seems to indicate that the stabilization would have to occur on such an early developmental level that many of the human qualities that enter into human happiness are lacking, as well as possibly the qualities which enable a person to achieve an insight into himself. The general emotional level lacks heights and satisfactions lack depth. In addition there are, of course, the many external criticisms possible of his "morality" in the light of social goals and standards, and the limited character of social progress possible on a pure expediency basis.

But perhaps the open character of the issue before us is in danger of being overrated. The Platonic tyrant had a dash of bravado about him, in spite of Plato's indictment. The Nietzschean superman is provided with a halo by being viewed as a bridge to the evolutionary advancement of the race. But the psychopathic personality even on the basis of present descriptions, whatever the interpretation, turns out under the miscroscope to be a pretty pitiful fellow. He is no fit support for the ethics of a carefree moral relativism.

§ Psychological Search
for Phenomenal Invariants

TRADITIONAL MECHANISTIC ASSUMPTIONS. The readiness to argue "Suppose there were . . ." to prove that "it all depends . . ." itself is significant. It reflects the traditional mechanistic assumption noted above that human beings are a neutral plastic stuff to be shaped by whatever forces impinge upon them. Contemporary psychology, with its conception of fundamental needs, has considerably weakened this traditional prop of ethical relativity.

Suppositions about what might happen are not the crux of the problem. True enough, if a man could be turned into a tiger or a moth, he would have different values. But obviously this settles no ethical issues. Nor does the mere range of beings into which men have actually been turned —from abject slave to creative spirit—of itself establish the correctness or incorrectness of the relativist outlook. Decision can come only from full analysis of the phenomena showing, on the psychological side, to what extent these states gave expression to man's fundamental needs.

DUNCKER'S ANALYSIS OF ETHICAL RELATIVITY. A quite different approach to a critique of mechanism in relation to values—different, that is, from a critique in terms of psychological needs—has been developed under the influence of Gestalt theories. A Gestalt view denies that man is a passive recipient reacting to discrete individual stimuli which, if repeated sufficiently, condition the mode of his response. For the Gestaltist, the response depends on the meaning of the situation, the whole configuration characterizing the object of awareness. Perhaps the most notable attempt to apply this to the problem of ethical relativity is to be found in an article by Karl Duncker entitled "Ethical Relativity? (An Inquiry into the Psychology of Ethics)."[48]

Duncker formulates the thesis of ethical relativity as follows: "There is according to this thesis of ethical relativity nothing invariable within the psychological content of morality. Any conceivable behavior may, in appropriate historical or ethnological circumstances, take its turn in fulfilling the function of social expediency."[49] Ethical relativity is thus construed as a thesis about the contents of the phenomenal or phenomenological field, not about its causal relations or functional relations. It may very well be true that morals serves the sociological function of securing the preservation

---

48. In *Mind*, New Series, Vol. 48, Jan. 1939. See also, Solomon E. Asch, *Social Psychology* (New York, Prentice-Hall, 1952), Ch. 13.
49. *Ibid.*, p. 38.

and welfare of the group, Duncker allows, but this is not the sole invariant of morals. On the contrary, he looks for a strict invariance within the phenomenal field itself.

HIS COUNTER-THESIS. Duncker's counter-thesis falls into several parts. First he establishes a concept of the "pattern of situational meanings" as distinguished from the act and the motive. The same act or motive can have different meanings: for example, usury changes historically from loans for consumption in a landed economy to loans for profitable production, as commerce grows. If we penetrate in each case to the pattern of situational meanings, the relativity disappears; for that is what the moral essence of an act, its ethical quality and value, depends on. Thus in western history usury is not first disapproved of and then approved of: "we have not two different ethical valuations of usury, but two different meanings of money-lending each of which receives its specific valuation."[50] Similarly, infanticide is not just an act with a single meaning. We have to see whether the infant is conceived of "as an inanimate thing, a limb of the group, a piece of property belonging to the pater familias, a person of its own, or finally as an immortal soul (which would be doomed to everlasting perdition if it died unbaptised)."[51]

The thesis of ethical invariance asserts that "given the same situational meanings an act is likely to receive the same ethical valuation. If an act is found to receive different valuations at different times or places, this is generally found to be due to different meanings."[52] For there are "general 'inner laws' of ethical valuation, the independent variables of which are meanings. Only within the range of these common laws is an ethical understanding of foreign ways possible."[53] So firmly convinced is Duncker of the role of meanings as independent variables that he offers the educational advice that one can compel a pupil's valuation by instilling

50. *Ibid.*, p. 41.
51. *Ibid.*, p. 43.
52. *Ibid.*, p. 44.
53. *Ibid.*, p. 50.

in him the emotional and cognitive meanings of a given virtue.

CONTRIBUTIONS OF THE GESTALT APPROACH. I have quoted and paraphrased at length because of the growing importance of this approach in contemporary ethics. For example, Köhler in his *The Place of Value in a World of Facts* develops the concept of "requiredness" as a kind of direct characteristic of a situation in the phenomenological field, and argues at length for its objectivity. It is essential, therefore, to distinguish what seems established and what seems dubious in the whole approach.

The central contribution of this Gestalt emphasis is its reinstatement of the phenomenal field—that is, the direct object of awareness—as an area for careful scientific scrutiny. This was done successfully on a large scale in the study of perception. When the visual field is described in terms of figure and ground, when it is described as containing an open or a closed figure, when one line is visually longer than another, and so on, there is no direct reference to the physical object on the one hand or the psychological act of perceiving on the other. The phenomenal field is explored for whatever facts and laws are discoverable within it. It provides an independent variable. Thereafter, fruitful correlations of an empirical kind may be found relating its contents to physical states and to psychological acts. The careful initial separation of areas for exploration pays dividends in the subsequent increase of knowledge. Nor should there be any special assumption in this procedure about the "status" of the phenomenal field or its reality. For reality judgments concern the relation of the phenomenal field to the physical and psychological. Thus one can describe the capers of ghosts that one sees, and then show their unreality by correlating the data of the visual field with physical movements in the environment and psychological fears based on traditional stories.

There is great scientific value in the application of such descriptive techniques to the moral field. It aids precision and in fact it furnishes a great part of the data. It was a comparable type of investigation which distinguished in moral judgments between being afraid, being ashamed, feeling guilty. In the psychoanalytic materials dealt with above, the correlation of feelings or emotions with dynamic conditions was central to the enterprise. It was such an approach which, in our argument with the stubborn man about ultimate disagreement of attitude, distinguished between a spoils attitude where opponents fought for the same good, a yielding to temptation where one side is acknowledged to be right by the other which nevertheless fights on, and a principled conflict where each regards the other as wrong; the mere behavior of fighting would not alone reveal the difference. Again, in discussing the bio-psychic propositions that "Life is a good" and "Pain is an evil" we used a similar technique. It was not enough that A killed or exploited or inflicted pain on B. We wanted to know what his attitude to B was, that is, what B was in his phenomenal field. Thus we could distinguish different steps to the brotherhood of man rather than treat it as an all or none concept. At the other extreme from brotherhood is the sense of the outsider as threat or enemy, a something-to-be-killed. We can learn a little about this from the "situational meaning" among some primitive peoples who do treat strangers as enemies. Then there is the Aristotelian attitude to the slave as simply a living tool, a means to the owner's goals. Beyond this there is the recognition of another as one who has values, without any necessary feeling toward him. There may be many further shades. The other person may be seen as one whose values may be my values (a possible partner). He may become a *fellow-man*. He may be one-of-my-group. He may be a brother (but now we must ask for the "situational meaning" of "brother"). These different forms a human being may take in another's eyes can no doubt be correlated with changes of circumstances in the growth of material conditions, in the consolidation of groups, the unification of the globe, and so on.

EVALUATION OF THE THESIS OF ETHICAL INVARIANCE. The

scientific value of careful inspection of the phenomenal field
need not, however, carry with it the thesis of ethical invariance in the phenomenal field. There is, in fact, a fundamental vagueness in the theory as Duncker presents it. For
it would seem to follow that every actual judgment of value
by a person accurately describing a value which he holds
is correct. If I disagree, for example, with a sincere disciple
of Hitler's who judges that Jews ought to be exterminated,
is it because I have not properly grasped his situational
meaning? If I read *his* description of Jews will I decide that
anything in fact corresponding to his vision merits destruction, but deny that the actual six million destroyed fitted
the description? Certainly more evidence is required to warrant such generosity to any and every conceivable value
judgment. What is more—if the thesis really implies this—it
propounds a more far-reaching relativity than it set out to
attack. It has literally, then, turned into its opposite.

Actually, Duncker presents little evidence for his thesis
beyond the few examples of usury, infanticide, modesty. He
has shown that some of what has appeared to be ethical
difference is not or may not be such. He has provided a
technique for diminishing indeterminacy in judgment, but
given little evidence how far it extends successfully. Evidence would have to cover the whole range of major social
value differences. For example, in a large-scale strike, what
evidence is there that the employer and the employee do
not see the situation in the same way—as a struggle for a
greater share of the products of industry—and yet differ on
the question of value? Similarly universal historical scope
cannot be secured by such devices as he takes from T. H.
Green's view that "It is not the sense of duty to a neighbor,
but the practical answer to the question Who is my neighbor? that has varied."[54] For this simply shows that the indeterminacy can be moved from the law to the operational
definition. To use this as evidence for his thesis Duncker

---

54. *Ibid.*, p. 48.

would have to show that either the emotional quality or the actual duties of a man to his "neighbor" have remained the same as the answer who was the neighbor changed. That is, he would have to establish either a proposition in psychology or in history or both.

The issue of the thesis of ethical invariance need not, however, be all-or-none. We may replace the thesis instead with the enterprise of searching for whatever invariances there may be discoverable in human value judgment. And some there very well may be. For example, a good case could be made for sympathy as such a value. The hypothesis is that whoever recognizes sympathy values it, and whoever does not value it will turn out on examination to be incapable of recognizing it. He will see it as pity instead. This hypothesis, although difficult to establish, is an empirical one and, whatever its truth-status, illustrates well the way in which an empirical meaning can be given to a specific "moral blindness" analogous to a particular color-blindness.[55]

The chief caution required in applying Gestalt techniques to values is not to stop short with the purely phenomenal. In the field of morality this is more important perhaps than in perception. For the aim of morality is conduct or action. And action does not depend purely on the qualities of the phenomenal field, but on correlating them with other situations in which the qualities of the field may change, and with knowledge of causes and consequences. For example, a man sees a sinister figure. Granted that the sinister quality characterizes the field and has phenomenal objectivity, the man does not rush to call the police. He considers the possibility that shadows or fears may be the cause of the field in which the quality appears. He calls on other experience.

---

55. I am indebted for this suggestion to Tamara Dembo who has made extended empirical and phenomenological studies of sympathy.

§ Limitations in the Psychological Treatment of Ethics

EXTERNAL EVALUATION OF PSYCHOLOGICAL VALUES. Clearly contemporary psychology, in its contributions to ethics, has covered a very wide field. For this reason we must, in conclusion, remember that it is not the whole story. This is especially important because there has been a strong contemporary tendency to offer psychological hypotheses to explain major social and historical phenomena.

It is obvious, to begin with, that central values in the psychological perspective may conceivably be challenged on extra-psychological grounds. For example, the religious tradition in some of its organized forms, such as Catholicism, looks on the mass of men as requiring authoritarian guidance in a paternalistic way. On such an assumption there might be a temptation to advocate full psychological maturity only for the few who will bear the burden of leading, but to maintain certain marked immaturities in the dependent masses. Defenders of this position might even recall that Freud, while he speaks of religion as "the universal obsessional neurosis of humanity," himself remarks that "the true believer is in a high degree protected against the danger of certain neurotic afflictions; by accepting the universal neurosis he is spared the task of forming a personal neurosis."[56] Why propose an ideal, they might add, striving towards which among the realities of life will produce a fundamental disillusion and despair? Will not the alternative proposal make for the least suffering in this sad world?

The resolution of such an argument, posing an external evaluation of the psychological ideal of maturity in terms of its results in the world, cannot be a psychological matter alone. It depends upon the character of the world—social, historical, biological, metaphysical.

---

56. Sigmund Freud, *The Future of an Illusion,* translated by W. D. Robson-Scott (London, Horace Liveright and the Institute of Psychoanalysis, 1928), pp. 76-77.

SOCIAL AND GROUP VARIABLES UNDETERMINED BY PSYCHO-LOGICAL VALUES.   Even within the limits of unquestioned psychological values, there are further social and group variables which remain undetermined by the psychological perspective. Any psychological criterion may be compatible with a variety of social forms. Thus hunger can be satisfied with a variety of acceptable food patterns. Similarly, a critique of an institution carried through in psychological terms may be perfectly sound without being decisive. For example, a competitive economy may receive a negative evaluation for the intensified aggressiveness it produces and the higher incidence of neuroticism and frustration. But whether there are compensating social gains depends on the social comparison of this and alternative economic institutions. The value of the psychological critique is not diminished by seeing it as less than the total basis of judgment.

Again, and perhaps most important, there is the simple fact that in purely psychological terms it is left indeterminate whether one group may or may not seek to satisfy its psychological needs at the expense of another group. A psychological determination of this ethical issue is not inconceivable on the pure level of speculation. It would require even more than Fromm's thesis that only if a man loves himself can he really love anyone else. To turn the trick it would require that no man could really love himself and be emotionally secure unless he loved all mankind. But for this extreme thesis there is little evidence. It is much more likely that the ethical issue of group relations requires a social and historical resolution.

WILL PSYCHOLOGY OUTGROW ITS INDIVIDUAL-CENTERED-NESS?   It is quite possible that the several shifts in psychology we have traced as altering the picture of ethical relativity themselves reflect a basic change in the psychological perspective. As I see it, there is a central theoretical shift stemming from an attempt to outgrow the assumption of the self-enclosed individual. As distinguished from the group emphasis in evolutionary biology, psychology has been predominantly individual-oriented. Its successes have come from the intensity with which it has explored the individual, and its shortcomings from concentrating almost wholly upon him. Contemporary psychology has emancipated itself from an outlook that identified the individual mind with its states of consciousness. But even in looking for physical correlates or relating environmental stimuli and individual response, it has maintained an individualistic model of explanation. When the model was extended inward into the individual the needs and forces discovered or assumed were given an egoistic cast. When it was extended outward towards the group, social psychology looked for individual ties to other individuals in the form of imitative reactions or libidinal attachments. The society of others has been regarded as a causal agent in determining response or as an instrumental agent in furnishing satisfaction. But it has not yet been given its adequate place as the content of the field of which the individual is a part. That the social relationship is part of the very texture of individual life, that it is a quality and not merely a condition of living, that a legitimate scientific concentration on the individual does not entail an individualistic model in understanding him, that the social and cultural are integral to and not simply added upon the psychological—all these are lessons with which the growth of psychology in its present direction may very soon present us. If this brings to fruition the fuller understanding of the old maxim that man is a social animal, it also points to the area of the social sciences as one in which ethical indeterminacy may be further narrowed.

VII. Cultural and
Social Perspectives

In the relations of ethical theory and social science ethical relativity has been rooted in a syllogism which may be loosely expressed as follows:

> Ethics is a cultural phenomenon,
> Culture is relative,
> Therefore ethics is relative.

We shall have to reckon with this inference, that is with the cultural and social content of moral ideas, the meaning of cultural relativity and the evidence concerning its theses, and the net result of this estimation for the issue of ethical relativity. Thereafter we shall turn directly to those avenues of socio-cultural investigation that give fair promise of lessening indeterminacy in ethical judgment. Finally, we shall look at the implications of our analysis for the familiar controversy as to whether scientists can make value judgments.

§ Cultural and Social Content
of Moral Ideas

MORALS A PART OF CULTURE. The social sciences, especially anthropology and history, have shown—one might almost say beyond cavil—that moral patterns are part of culture, intimately related to the life that is going on. The specific types of relations or ways in which moral ideas depend on the rest of the cultural and social milieu require careful analysis.[1] But the fact of the dependence can be seen in a preliminary fashion by noting the subject-matter with which moral ideas are usually concerned and the extreme difficulty we find in understanding a moral idea without looking to specific socio-cultural content.

If we take our own moral pattern as an example, there are many rules concerned with property relations—such as prohibitions against stealing, or coveting others' goods, as well as a large part of the ideal of justice. There are also many rules and moral attitudes on sex and the family which govern procreation and the care of the young. Rules against killing within the group, for obedience to law and constituted authority, for mutual assistance, are parts of a pattern of social control, required for maintaining group life and preserving the group itself. Beside all these rules, there are all sorts of attitudes considered desirable in ordinary interpersonal relations (courtesy, sympathy, respect, and so on), ends (such as success and achievement), and virtues (self-control, prudence, the various intellectual virtues, courage) which set the socially approved goals for individual striving and regulate ways of handling oneself in the world of man and nature.

Many of these rules, attitudes and character-traits sound very abstract, but they have a quite specific socio-cultural content. It is difficult to understand the meaning of justice without looking to the modes of distribution of gains and burdens; these vary in different societies. The meaning of chastity has reference to marital patterns; and honoring one's father and mother may be enjoining a pattern of religious ceremonial, economic support, political obedience, or personal affection, or some configuration of several of these elements.

THE SEARCH FOR "ESSENCES." The chief dissenting voice to the view that moral ideas are socio-cultural phenomena

---

1. This is carried out in part below, pp. 231 ff.

comes from the major tradition in philosophical theory which looks for moral "essences." It wants to know what respect for parents, justice, chastity, success are "in themselves," and looks upon socio-cultural references as merely "applications" of an antecedently determined and logically independent idea. Its effort is to distil the materials until the pure essences can be extracted and held up to shine in the light of intellect or bask in the glow of sensibility. Thus honoring parents becomes a pure idea of piety, justice is rarified into a formula of "to each his own," chastity is transformed into purity and integrity, and success becomes achievement or the actualization of capacity and ends up as self-realization.

If, in the light of the methodological discussions in previous chapters,[2] we discount the metaphysical aspects of this search for essences, we may interpret it as primarily the search for invariants of a psychological type in different socio-cultural contexts, including the search for phenomenological constants.

Such an inquiry may sometimes be successful. There may, for example, be a common psychological quality in all child-parent feeling irrespective of variety of form of social obligations. But on the whole it is far more likely that if we are dealing with separate elements at all, they do not form "mixtures" in which the psychological component maintains separate properties, but "compounds" in which the elements are transformed. Certainly in the case of success, we have no a priori right to equate the self-satisfaction of a craftsman looking at his completed work with the feeling of a man who has made a fortune in a commercially-minded society. The very structure of their feelings may be quite different. To equate them both as "self-realization" is either the obvious assertion that values have in both cases been actualized —which would make almost any values alike—or else the far-reaching empirical hypothesis that both play similar functional roles in a process of psychological development.

---

2. Especially of the absolutistic views of conscience, reason and moral law in Chapter 2, and of first principles and definition in Chapter 3.

Similarly, in the case of phenomenological constants we have already seen that the attempt to find them may lead to a fuller social differentiation of the content rather than a transcending of the social.[3]

Where there are no such psychological invariants or phenomenological constants, we seem to be left with a very bare formula of a highly abstract type. But this again turns out on analysis to have a social dimension in its meaning. Thus justice may be stripped of specific distribution patterns and specific psychological elements such as the feeling of unfairness. The bare remaining idea of "to each his own" is then simply the idea of a possible system of individuals (implicit in "each") related by some pattern of apportionment ("his own"). So far from the pure idea being logically prior to the social system, the meaning of the idea presupposes the idea of a social system. The further proposition that justice is a universal idea thus reflects the sociological generalization that every society in fact finds it necessary to establish some pattern of apportionment among individuals.

§ The Meaning of Cultural Relativity

VAGUENESS OF THE CONCEPT OF CULTURAL RELATIVITY. If moral ideas are in a significant sense predominantly sociocultural, we must next inquire concerning the second premise, that culture is relative. Now in spite of its wide use, the concept of cultural relativity lacks precision.

The basic phenomenon to which it points—just as in the thesis of ethical relativity—is variety and difference in human behavior in all its aspects in different societies. The older cultural relativists rested their case on the comparison of specific items in various cultures. With the growth of pattern analysis it was seen that not merely items but whole configurations varied.[4] This configurational variety pene-

3. See Duncker's discussion of usury, above p. 194.
4. This lesson was elaborated in Ruth Benedict's well-known *Patterns of Culture* (New York, Houghton Mifflin, 1934. Reprinted by Penguin Books, 1946.)

trated deeply into the particular acts, qualities of feeling, even modes of thought of the different peoples. Thus Mead and Bateson see that the fact that the Balinese in picking up a pin use only the muscles immediately relevant, instead of involving almost every muscle as an American or New Guinea native will, as having deep roots in the whole culturally developed character structure.[5] The fact that the Kwakiutl Indians feel outraged or insulted by the death of a member of the in-group is not merely a remarkable difference in reaction but is integrated with their whole cultural mode of interpersonal relations.[6] The very concepts in which one culture casts its understanding of the world may differ markedly and perhaps untranslatably from those of another, so that the systematic picture of reality will be culturally fashioned in some different way.[7] Relying on such data the cultural relativist today assumes with confidence the pervasive scope of differences in human cultures.

If we look, however, to the meaning of the doctrine of cultural relativity as an interpretation of this diversity, we find several separate components. Some focus on the problem of explaining human behavior, others on value phenomena specifically, and there even takes place occasionally the conversion of the doctrine from an explanatory thesis to a value outlook.

INSISTENCE ON A GENERAL-CULTURAL VARIABLE. The wider minimal meaning is seen in the old battle in the social sciences against reductionism, the familiar struggle of nurture vs. nature. Here the insistence that human behavior is relative to culture constitutes the use of a *general-culture* variable. It is the claim that a category of culture is unavoidable, that purely biological, ecological, psychological categories are insufficient for an explanation of what a man does

---

5. Gregory Bateson and Margaret Mead, *Balinese Character* (Special Publications of the New York Academy of Sciences, Vol. II, 1942), p. 17.

6. Ruth Benedict, *op. cit.*, Ch. VI.

7. See, for example, Dorothy Lee's analysis of Trobriand concepts in "Being and Value in a Primitive Culture," *The Journal of Philosophy,* XLVI, 401-415 (June 23, 1949).

and how he behaves, that there is a learned dimension in human behavior.

INSISTENCE ON UNIQUE PARTICULAR-CULTURE VARIABLES. The differentiating thesis of cultural relativity is to be found beyond this in an insistence on the uniqueness of particular cultures. It is held that every culture has its own laws or structure and is a kind of universe by itself. Inherent in this doctrine is the belief that no significant laws can be found for all societies—for one thing, because no term offered for such laws would mean quite the same thing in referring to one culture as it did in referring to another. Cultures are not, then, like different planets whose position can be predicted from the previous state of the system by means of common laws of planetary movement. Nor are they even like different species, whose own "laws" may be seen in the more systematic framework of an evolutionary development. They are unique particulars which can be analyzed and appreciated but not compared.

It has often been the fashion to ascribe this alleged separateness and uniqueness of culture to the nature of the human spirit, as contrasted with matter in the natural world whose uniformities science studies. The social sciences in this view are more art than science, painting portraits of cultures in all their individual detail. Sometimes the underlying assumption here is a philosophical idealism, as in Spengler's thesis that every historical culture has its own underlying idea.[8] Perhaps the greater number of anthropologists and sociologists today, however, no longer think in such terms as remove culture from the natural world. Their interpretation of cultural relativity tends to drop metaphysical or a priori bases and appeals instead to complexity, the role of historical accident, the great variety of detail, which make generalization difficult. Such factors are very similar to the ones we found in looking for the sources of

8. Not every philosophical idealism yields this consequence; Hegel propounds historical laws that transcend any given culture, even though he is one of the fathers of the theory that every culture has an inner pattern.

indeterminacy in ethical judgment—the primary sources of field complexity and field instability.[9] They do not provide a hardened prohibition of going beyond particular cultures but only a sad warning that such attempts may fail. But they should include the realization that pessimism may be reflecting the present state of knowledge rather than an ultimate character of the field.

CULTURAL RELATIVITY AS A VALUE THESIS. If it is claimed that cultures are unique and particular in all their aspects, then obviously values, morals, ethics, as part of culture, have the same character. This entails in the first place the recognition of diversity. We still speak of moral feelings or feelings of conscience, but we can no longer have the happy belief that this is a single determinate phenomenon within the human breast, as Bishop Butler expounded it in the 18th century. Instead, we find a vast array of different blends, with guilt predominating in some, shame in others, simple fear in others. The organization of particular values will take different shape: there are success-bent cultures, competitive cultures, cooperative cultures, fear-dominated cultures, and so on. And in the very procedures of evaluation there will be variety: there are authoritarian theocratic cultures, mystic individualist cultures, rationalistic cultures, and so on. The cultural relativity thesis would assert the ultimacy of these differences, their irreducibility, and perhaps even the non-translatability of their conceptual components.

An even wider thesis is sometimes found on the current scene. This is due to a special trend in anthropology and sociology which describes the very unity of a culture in *value* terms. Benedict's *Patterns of Culture,* for example, seeks a common value outlook carried in the prevalent institutions and attitudes of the culture and transmitted through educational and social experience from generation to generation. In such an approach practically every phase of life activity of a society is seen as a value-selection out

---

9. See above, pp. 96-100.

of possible alternatives. An individual's physical habits, interpersonal attitudes, as well as goals and moral outlook, express a type of personality molded by the society into a value-pattern. In such a view cultural relativity and ethical relativity seem to merge, if the unique particular-culture variable is insisted on. It is not so much that cultures vary and with them their values as that the essential difference between cultures is already established as irreducibly different value-premises.

In either form—whether as a special application or as an expanded account of cultural selection of patterns—the value thesis of cultural relativity finds it easy to proclaim that science proves there are no universal goods to be defined by universal strivings, universal processes, universal contexts. It has given a much richer meaning to the older contention that there are no trans-cultural standards because in Sumner's words, "the standards of good and right are in the mores." Any trans-cultural ethical comparison is therefore a spurious pursuit, an attempt to foist our values or the values in our methods of inquiry on other peoples.

Sometimes also the thesis of cultural relativity is presented as if it were a philosophical outlook, itself incorporating definite values. For example, Herskovits characterizes cultural relativism as "a philosophy which, in recognizing the values set up by every society to guide its own life, lays stress on the dignity inherent in every body of custom, and on the need for tolerance of conventions though they may differ from one's own."[10] Such a formulation, we shall see, borders on an attempt to assert an ethical absolutism on the basis of a cultural relativism.

§ Critique of the Theses
   of Cultural Relativity

NECESSITY FOR A GENERAL-CULTURE VARIABLE. That a cultural variable is required in explaining human behavior

10. Melville J. Herskovits, *Man and His Works* (New York, Knopf, 1951), p. 76.

is an established lesson of social science. This minimal thesis is coextensive with the knowledge of the full scope of cultural influence in fashioning men's ways. But its recognition should not carry with it any thesis of the limitless powers of culture or of the limitless plasticity of man. An over-sharp dualism between the biological and the cultural, combined with the ascendancy of culture in the controversy has sometimes led to the too ready assumption that there are no universal tendencies or strivings in behavior which do not yield to cultural influences. It has been sometimes assumed that any drives, whether biological or psychological, are wholly shaped by the social pressures of culture and that men can come to want anything their culture teaches them, feel anything their culture expects of them, and behave as their culture stamps them. Only such an extreme assumption explains why so many theorists have felt on reaffirming the existence of basic human needs, that cultural relativity was automatically refuted. A contemporary reckoning with this issue will have to break through the underlying dualism and see the full extent of the interpenetration of the cultural with the biological and psychological at every step.

UNIQUE PARTICULAR-CULTURE VARIABLE NOT ALWAYS REQUIRED. The more specific thesis of cultural relativity, that all phases of human life require in their explanation a reference to a unique particular culture in some ultimate and irreducible way, appears incorrect in this extreme form. Some phenomena although expressed in the mode of a particular culture turn out to be trans-cultural, and some phenomena although very specifically wedded to a particular culture are capable of trans-cultural systematization. This does not of course deny the vast amount of cultural material that may be limited to a particular-culture reference.

Scientific knowledge is the best example of an area that has withstood repeated attempts to submit it to cultural relativization. There is no inherently Eskimo physics, though there may be Eskimo discoveries in physics, or a uniquely Eskimo garb to physical concepts cast in supernaturalist terms. This does not then deny that particular cultural elements frequently cling to models that are employed, especially in the social sciences. But the history of science shows that when this is discovered, the discovery, so far from establishing the inevitable dependence of science on a particular cultural view, provides instead concrete tools for removing the reference and refining the scientific ideas.[11] Similarly the fact that there will always be a cultural element in the mode of expression and communication takes second place to the possibilities of intercultural scientific translation. Contemporary anthropological analyses of language and thought in different cultures need not underestimate the extent to which there is all-human recognition of a common world.

There are a growing number of areas in which phenomena strikingly singular in their particular culture reference may become systematized in a psychological or inter-cultural framework. For example, witchcraft in some primitive societies may fall in the same body of systematic psychological theory as scapegoating in "advanced" cultures. And any successful systematic theory of social structure might readily show a diversity of particular elements performing comparable functions. Thus judgments of ultimate irreducibility, uniqueness or plurality may in some areas turn out to be reflecting an inadequate state of knowledge.

VALUE THESIS OF CULTURAL RELATIVITY ACCORDINGLY UNSETTLED. The denial that ultimate irreducible particular-culture reference is inevitable throws every area of human life open to independent empirical investigation to see how culturally relative it in fact is. With the second premise of our initial syllogism thrown into such doubt, the conclusion cannot rest on a purely deductive backing. The inquiry into values, morals, ethics and the extent of indeterminacy there existing becomes part of the very evidence for or against

11. This same point was made above in connection with attempts to discredit reason in relation to psychological and sociological analyses of knowledge. See pp. 52-53. For fuller discussion, see below, pp. 282-289.

the scope of the cultural relativity thesis. And this is equally true of any of the expanded value theses which regard all cultural differentiation as a kind of value-selection.[12]

INCONSISTENCY IN CONVERTING CULTURAL RELATIVITY INTO A VALUE OUTLOOK. A formulation which insists upon an irreducible particular-culture variable in explaining all phases of human life finds itself in unavoidable difficulty if it asserts a universal value outlook.[13] For if the latter is accepted it is evidence against the former thesis. And if the latter is given a relative cultural status, it becomes a local or provincial outlook. Thus where the appeal is made for the dignity of every body of custom and for cultural toler-ance as universal values, these values must either be estab-lished on some absolute basis, or else found to be empirical invariants. (For example, it might be found that in spite of differences in material conditions, every known cultural group in fact possesses some degree of dignity and some appreciation of it.) In either case the establishment makes a breach in the case for a complete cultural relativity, in which the meaning of dignity in the different cultures would be incommensurable. On the other hand, the plea for dignity and tolerance as values may be simply a reflection of a culturally developed liberalism in the western world.

That different moral lessons may in fact be associated with the recognition of the central role of culture in fash-ioning value differences is clear on comparative grounds alone. For example, Sumner's assertion that the mores can make anything right and prevent the condemnation of any-

---

12. Such broad use of the value concept raises, however, further issues. See my "Concept of Values in Contemporary Philosophical Value Theory," *Philosophy of Science*, 20, 198-207 (July 1953).

13. For a fuller critique of this position, see David Bidney's analysis of cultural relativism and the transvaluation of values in his "The Concept of Value in Modern Anthropology," in *Anthropology Today*, ed. A. L. Kroeber (Chicago, University of Chicago Press, 1953), pp. 682-699. Con-siderable recognition of the unstable theoretical situation in this area is found in the collaborative assessment by anthropologists themselves of the current status of their problems: see the discussion of values in *An Ap-praisal of Anthropology Today*, ed. Sol Tax and others (Chicago, University of Chicago Press, 1953).

thing is scarcely presented in such a fashion as to inspire one with a sense of the dignity of different ways. It is tempting to see it instead as an expression of his view that the masses are passive. Hofstadter epitomizes his outlook: "Like some latter-day Calvin, he came to preach the predestination of the social order and the salvation of the economically elect through the survival of the fittest."[14] In fact, Sumner's analysis of the relation of mores to the interests of the society leads him to a practical abandonment of relativism. For example, speaking of eugenics he says: "This much, however, is certain,—the interests of society are more at stake in these things than in anything else. All other projects of reform and amelioration are trivial compared with the interests which lie in the propagation of the species, if those can be so treated as to breed out predispositions to evils of body and mind, and to breed in vigor of mind and body."[15] And again, on the topic of inbreeding, "We who have inherited the taboo now have knowledge which gives a rational and expedient reason for it."[16] What more is required for trans-cultural bases of evaluation than that societies have interests which are basic, that there be definite goods and evils of mind and body, that there be rational grounds for adoption of policies, and that different modes of organizing society can be compared on such bases? Irrespective of the content, Sumner seems here to grant that there are conditions for a rational morality. If so, then why cannot a society become conscious of them and guide its morality in these terms? The "right" and "wrong" which can be made merely by mores become simply moral beliefs, not moral realities.

On the other hand, Ruth Benedict's analysis of culture patterns combines a strain of implicit cultural relativity with an affirmation of several values, overarched by those of dignity and tolerance we have already considered. Thus

---

14. Richard Hofstadter, *Social Darwinism in American Thought 1860-1915* (Philadelphia, University of Pennsylvania Press, 1945), p. 51.

15. *Folkways* (Boston, etc., Ginn, 1934), § 532.

16. *Iibd.*, § 533.

Elgin Williams in a value analysis of *Patterns of Culture* finds Benedict implicitly condemning war, using absence of violence as a value criterion in a culture's handling of sex relations, condemning initiation ordeals which are part of an authoritarian technique, and warning against despising the body.[17] He finds this in conflict with the general relativistic tolerance of the book as expressed in such statements as: ". . . Adequate social orders can be built indiscriminately upon a great variety of foundations . . . ," ". . . Ends and . . . means in one society cannot be judged in terms of those of another society, because essentially they are incommensurable," "No man can thoroughly participate in any culture unless he has been brought up and has lived according to its forms."[18] And finally, he points to Benedict's attempt to make us more self-conscious about our own ways, so that we may appraise them objectively. He finds this whole phase of her work out of keeping with her claim of "coexisting and equally valid patterns of life which mankind has created for itself from the raw materials of existence."

FOUR AVENUES OF INQUIRY ON ETHICAL RELATIVITY IN THE SOCIO-CULTURAL SPHERE. We may conclude that none of the theses of cultural relativity furnishes a prior determination of the extent to which ethical judgment may be rendered more determinate on cultural and social bases. On the contrary specific inquiry in this direction may furnish evidence or throw light on the limits of cultural relativity itself.

Four avenues of research into this further inquiry may be suggested. They concern the types of order, or invariants in the cultural and social domain which, if established, might serve as bases for trans-cultural evaluation. A starting-point might be found in *invariant values,* either empirically discoverable on the surface by the comparison of cultures or else revealed by stripping off analytically differences that are irrelevant. If there are such perennial values, and if

17. Elgin Williams, "Anthropology for the Common Man," *American Anthropologist,* New Series, 49, 84-90 (January-March 1947).
18. *Ibid.,* p. 88.

there is no other ground (for example, in terms of psychological or biological valuations) for their rejection, they may provide a basis for assessing social and cultural forms that express or impede their realization. A second type of order is found in the direct application of lessons already established from the biological and psychological perspectives, using the criteria of survival and well-being, of physical and mental health. A third avenue to possible order lies in the invariant tasks that every society must undertake; these may provide structural criteria for measuring success. Finally, it is also possible that knowledge of specific sociological determinants and functions may provide a basis for trans-cultural judgment. Even where the situation differed in various societies, it might be shown how the same value was supported by different cultural mechanisms and through different cultural forms under differing conditions. In general, it is to be noted, trans-cultural ethical judgment need not mean the same specific values everywhere; it may mean that even without identical values, determinate evaluation of particular cases is nevertheless possible.

§ The Search for Invariant Values

REASONS FOR FAILURE HITHERTO. The first path—the search for significant invariant values—has, of course, so far not proved very fruitful; and this failure has been one of the pillars of ethical relativity. It is important, however, to indicate why it would be premature to abandon the search. The problems of evidence are extremely complicated. It must not, for example, be assumed that there is a disagreement in values simply because the same value has not grown up. The conditions may be different. That smoking is a widely valued pleasure among people with very different ways of life—setting aside the question of its own further evaluation in terms of health and psychological function—is apparent now that it has been introduced all around the world. This could hardly have been discovered when only Indians knew tobacco and its uses. Nor can we make aesthetic judgments on the evidence that the ancient Greeks did not take delight in movies—since they had none. Modern classical scholars could, no doubt, have quite an argument on the question of the probable aesthetic reactions of 5th century B.C. Athenians to Hollywood films; but this would not be evidence. Again, there are differences of belief on matters of fact. The illustration sometimes found of abandoning parents (or killing them) before the advent of physical decline—on the ground that a person forever remains in the hereafter in the state in which he departs this life—clearly is no evidence against filial piety; it may even be the strongest evidence for it. Nor do we really know the state of feeling in which the children performed this duty. To discount for distorting or preventive factors is also a difficult matter. We would not take as evidence against the joys of swimming, the adverse judgment of a man who could barely swim, or who learned to swim in old age and found it a strain. Similarly, we should discount the judgment of a people among whom swimming was tabu, just as we would not consult a blind man's opinion on the beauties of a sunset. In addition, one must distinguish between valuing something as a means and valuing it as an end; some alleged value differences may refer to means values. Thus Kluckhohn says: "No society has ever approved suffering as a good thing in itself—as a means to an end, yes; as punishment, as a means to the ends of society, yes. We don't have to rely upon supernatural revelation to discover that sexual access achieved through violence is bad."[19] When all the complex factors to be analyzed are taken into consideration, the search for invariants is not hopeless; it has scarcely begun.

PROMISING AREAS. Such studies might begin cautiously with areas of individual activity in which common structure and common processes suggest that there may be common values. Here belong studies in similarity of values in ele-

---

19. Clyde Kluckhohn, *Mirror for Man* (New York, Whittlesey House, McGraw-Hill, 1949), p. 285.

mentary perceptual or aesthetic response, in joys of bodily movement, in developmental process values such as independence and increase of immediate mastery of nature, in joys of cognitive and reflective processes, and so on. A second area, not wholly divorced from the first, concerns basic values in interpersonal relations, in association, mutual recognition and affection.

Beyond these lie the many areas of developed creative activity in human life—in social life, art, science, emotional response to complicated symbolic systems, and the whole range of interpersonal relations. A systematic canvass of invariant valuations has been limited hitherto by the readiness to accept negative votes without inquiring whether they were absolute rejections or cases of values outweighed by other considerations. It has also lacked in the past sufficient psychological equipment to deal with distorted perspectives or fear reactions. This whole general line of inquiry still remains open.

Nor should the possibility be excluded that even highly abstract values may be found invariant in human societies, although supported by extremely varied material; for example, that the very intensity of conflicts and frictions may make order and harmony universal though unrealized ideals, the extent of men's sufferings at each other's hands enshrine justice as an invariant hope, or the scope of recurrent problems make knowledge as the key to mastery of nature almost a supreme invariant.[20] MacBeath, who has carried out one of the broadest comparative ethical surveys in recent years, maintains not merely that human nature and its needs furnish a common raw material but also that there is a common formal ideal at which men aim, whether conceived as the good for man, the greatest good of the greatest number, the good of the self as a whole, and so forth.[21] In between are operative ideals articulating the formal ideal, finding institutional embodiment and guiding practice. He has not

20. See below, p. 261 f.
21. A. MacBeath, *Experiments in Living* (London, Macmillan, 1952), p. 68.

shown adequately that the formal ideal is a definite functioning value structure rather than a hypothetical construct of the theoretician resting on analogy. But an approach of this sort is suggestive for further inquiry.

INVARIANT EVILS. Comparable to the search for invariant positive values is that for invariant evils. On the whole these have stood out more clearly in their personal psychological forms—pain, frustration, etc.—than in specific social forms. Yet perhaps some social practices, ways of living, and group organization may turn out when carefully identified and delimited to be held universally as evil. Killing within the in-group probably has this character.[22] All societies in which witchcraft occurs seem to regard its exercise against people, where it is not for recognized punitive purposes or recovery of stolen goods, as evil; that is, in spite of differences in content, there may be a core of "black" magic beyond the pale of acceptance. Similarly universal tabus such as on incest, in spite of varying range of prohibition, contain a core of what is regarded as wrong. That judgments of invariance may be secured by refinement of content was suggested in Duncker's discussion of the phenomenal field.[23]

Considerable attention in contemporary sociology has been paid to the problem of *disorganization*.[24] Such a concept points to the structural conditions of recognized evil in a society and its analysis might be expected to articulate invariant negative social values. But on the whole so far it appears to have provided only a miscellaneous category for problems arising from spreading departure from established norms, impact of changing conditions on people adhering to traditional ways, social effects of the incidence of personal pathology or disintegration, and so on—a kind

22. Cf. Franz Boas, section on Ethics in article on "Anthropology," *Encyclopedia of the Social Sciences,* v. 2, pp. 97-98.

23. See above, pp. 193 ff.

24. For a brief presentation of selections giving different approaches, and a bibliography, see A. M. Lee and E. B. Lee, *Social Problems in America: A Source Book* (New York, Holt, 1949), pp. 20-35.

of general indication of sore spots or social tension areas. It is possible that a more general theory of invariant social evils may be advanced by more specialized study of specific aspects of sore spots. For example, there may be generalizations possible about types and limits of deviance permitted in a society, or the limits of permissible latitude in the selection of means for socially approved goals, or the rate at which change can be absorbed.[25] Such studies would have to rest on a comparison of types of social structures and on historically different types of material culture and economic organization.

PRESENT PRELIMINARY CHARACTER OF SUGGESTED INVARIANTS. At present, any discovery of perennial values or invariant evils growing in all cultures or having the given character in all societies in which they occur, would constitute only a beginning in our inquiry. One would still have to ask which of them are chance elements, which spring from the general conditions of human life, which derive their value from the satisfaction of biological and psychological needs, and to what degree the universal element is central or peripheral in the value configuration as it occurs in the actual lives of men in particular cultures. And, of course, the mere fact of universal positive valuation would not mean that the value was itself immune to evaluation in terms of its relations and role in human life.

---

25. Conceptual frameworks on problems of deviance are offered in Talcott Parsons, *The Social System* (Glencoe, Illinois, Free Press, 1951), Ch. VII, and Robin M. Williams, *American Society* (New York, Knopf, 1951), Ch. 10. On the problem of means, see the typology proposed in Robert K. Merton's "Social Structure and Anomie," in his *Social Theory and Social Structure* (Glencoe, Illinois, Free Press, 1949), Ch. IV. Theories of the rate of social change that can be absorbed have a range historically varying from the idealization of fixity to the assumption that any change can be absorbed for which material conditions are adequate. Current industrialization programs in countries under varying social systems (e.g., China and India) are focussing attention on such issues.

§ Application of Biological and Psychological Criteria

CRITERIA OF MENTAL HEALTH. The application of biological and psychological value criteria to the cultural domain —the second proposed path to trans-cultural evaluation— has made perhaps the greatest advances recently in the search for invariant elements. This is chiefly because these perspectives have undermined the old mechanistic view of man as almost wholly plastic material. The negative evaluation of frustration and break-down provides a minimal basis for trans-cultural evaluation. A brief paper by Ruth Benedict on this topic is an interesting addition to her outlook discussed above.[26] Benedict suggests as a universal cross-sectional criterion of mental ill health the loss of ability to go on functioning. In addition there are special criteria of mental ill health that can be added in the context of each culture. For example, suicide can be used in our culture, but not in Japan where the suicide of a samurai "was the final honorable act of a well-balanced man." Again, "An adult who strikes a child must be mentally ill in many societies, and the American Indians often comment on our usual chastisement of children as we would comment on an insane homicidal act."[27] The point is not whether suicide or corporal punishment for children are ethically desirable or acceptable; it is simply that ways can be worked out for identifying mental illness which transcend cultural differences, even though forms of expression vary in different cultures.

Benedict goes on to illustrate the attempt to correlate cultural institutions with mental health and ill-health. Using the degree to which the individual is singled out for humiliation before his fellows, she shows the wide range in which cultural ceremonies may be structured with respect to this

---

26. "Some Comparative Data on Culture and Personality with Reference to the Promotion of Mental Health," Publication No. 9 of the American Association for Advancement of Science, pp. 245-249.

27. *Ibid.*, p. 246.

property. At one extreme, puberty ceremonies, courtship, marriage, and so on, are gala occasions, at the other, humiliations and dependencies. In some societies, humiliation is associated with transgression, in others it is "a recurring life-experience from which no individual can defend himself."[28] The latter, again, are of two sorts: in one, some positive cultural arrangement for dealing with humiliation is available to the individual; in the other, there is none, and here is where there is greater incidence of breakdown. Such distinctions clearly provide a scale of trans-cultural comparison, evaluating social forms by the degree to which they foster or are inimical to mental health. Benedict suggests that similar studies could be made, in relation to mental health, for economic institutions which govern distribution of goods, familial institutions which determine the extent and sharpness of the kin group, etc.

What could such indications of social ill health be? Possibly one could try to develop a general concept of level of anxiety in a society. Some indices such as the frequency of alcoholism are promising in this direction. Horton, for example, uses the assumption that greater anxiety requires greater amounts of alcohol to reduce it, in a cross-cultural study of patterns of drinking behavior and their relation to release of sexual and aggressive impulse.[29] In any case incidence of alcoholism may point to some special area of tensions within the structure or conditions of the particular group. In primitive societies, Benedict once suggested, the prominence of witchcraft may also be a useful index of degree of social tensions.

If the use of such minimal criteria appears to open by no means narrow possibilities, even more may be expected from a trans-cultural application of the growing criteria considered above.[30] These were the biological criterion of phys-

---

28. *Ibid.*, p. 248.
29. Donald Horton, "The Functions of Alcohol in Primitive Societies," in *Personality, in Nature, Society and Culture*, ed. by Clyde Kluckhohn and Henry A. Murray (New York, Knopf, 1948), pp. 540-550.
30. P. 185 f.

ical health, the positive psychological criterion of the satisfaction of emotional needs, and the as yet undeveloped criterion of positive happiness. In part they were illustrated in their original presentation, but a few comments on each is pertinent here.

HEALTH AND LONGEVITY.   The criterion of health and longevity rests on the conclusion that these are goods for all people everywhere. Therefore institutions—technological, economic, religious, etc.—which impinge more or less directly on the productive processes of the society and on its health, shelter and sanitation arrangements, can be evaluated in this respect. This does not mean glorifying technical progress solely, nor exalting bathtubs over the spiritual life. A fear of this sometimes inhibits the use of these criteria. Another source of the inhibition is the desire to remove responsibility for the state of colonial peoples from the dominant industrial peoples, by the comforting belief that "they like dirt and disease and a short life-span, because it is part of their way of life and has always been."

Even where such suffering is tied to cultural institutions which in turn support positive values, it does not follow that the suffering is any the less evil. This can be seen in our own society, for example, in automobile accidents. When we read the statistics of a holiday week-end's toll on the highways, and are tempted to dismiss it with the remark "It's the price of an industrial society," it is well to correct the judgment in two ways. The first is to think of the individual case; the second to deliberate on what actually can be done—in highway building, in education, in the alteration of attitudes, in the planning of facilities—to minimize such losses. It is the same in other cultures. A low life expectancy and widespread disease are evils in any culture, no matter how far they may be rationalized in established beliefs that life does not count, that only eternity matters. The recognition of such universal evils does not make mere living the highest value. But there is all the difference in the world between a conscious choice, with full knowledge of the realities of the situation, to sacrifice life or health for a goal, and an escape myth that life does not matter or that greater glory will come from despising health and welcoming disease.

EMOTIONAL NEEDS. The satisfaction of emotional needs can become the basis, as we have seen in considering the psychological perspective, for the critique of whole constellations of character. It therefore provides a way of evaluating across cultures, especially modes of interpersonal relations, and institutions of child-rearing and education. Similarly, political and economic institutions can be evaluated insofar as they affect the quality of interpersonal relations, e.g., the extent to which and the conditions under which democracy enhances the individual's sense of dignity or competitive capitalism intensifies emotional insecurity. Especially today, in seeking western technology, non-western societies with long established cultural traditions find it necessary to ask how far the introduction of techniques to achieve greater industrial mastery will alter their particular ways of satisfying emotional needs and their often more highly developed patterns of interpersonal relations.[31] Evaluations on psychological grounds do not determine the whole field—for they will surely leave us with a wide range of morally acceptable ways—but they do constitute an important part of the ethical summation.

HAPPINESS. The criterion of positive happiness is, as suggested above, still insufficiently developed. Benedict indicated her belief[32] that the minimal negative criterion could be supplemented by a positive criterion of the happiness of a people, although this was a more difficult task; she did not, however, carry out this project before her death. There are indications in the anthropological and psychological lit-

---

31. For a suggestive discussion of these questions in the relations of America and India, see Gardner Murphy, *In the Minds of Men* (New York, Basic Books, 1953), esp. pp. 57, 286. Murphy suggests that there is much which we in turn can learn from India on the subtlety of interpersonal relations, and on such human qualities as simplicity, generosity, warmth, reverence, and especially reverence for the child. See also the recommendations for procedure in technological change, in *Cultural Patterns and Technical Change*, ed. M. Mead (UNESCO 1953), pp. 305 ff.

32. In conversation with the writer.

erature that such over-all judgments, cast as the synpotic picture of personality, are not lacking. An excellent illustration
is to be found in the conclusion of Bateson and Mead's study,
Balinese Character:

"In these various contexts of life the Balinese character is
revealed. It is a character based upon fear which, because it is
learned in the mother's arms, is a value as well as a threat. It is
a character curiously cut off from inter-personal relationships,
existing in a state of dreamy-relaxed disassociation, with occasional intervals of non-personal concentration—in trance, in
gambling, and in the practice of the arts.

". . . Always a little frightened of some undefined unknown,
always driven to fill the hours, so empty of inter-personal relations, with a rhythmic unattended industriousness, he follows
the routines laid down by calendars and the revelations of those
in trance, relaxed at the center of any world of which he knows
the outlines.

"Life is without climax, and not the ultimate goal but rather
the first impact of experience, the initial ping of startle, is the only
stimulus that has real power to arouse one's interest. And there is
always the danger that one may not be aroused at all. Between
the Death which is symbolized by the Witch's claws and the
graveyard orgies, and the death which is sleep into which one
retires when frightened, life is a rhythmic patterned unreality of
pleasant, significant movement, centered in one's own body to
which all emotion long ago withdrew."[33]

Now, apart from the question whether alternative developments are possible in the situation—clearly the process of
alteration of a living culture such as that of Bali is a complex
issue—does not this summary pass judgment that life is
being lived on the defensive? Can we then say that these
people are not as happy as human beings are psychologically
capable of being? Such over-all judgments require to be
examined more deliberately by the combined efforts of social
scientists and psychologists. It is purely a scientific question.
But the question itself is clearly far from the assumption

33. Gregory Bateson and Margaret Mead, *Balinese Character* (Special
Publications of the N. Y. Academy of Sciences, Vol. II, Dec. 7, 1942),
pp. 47-48.

of ethical relativity that each people has its own mode of happiness, that its ways define happiness for it, and cannot be judged from outside. And while one would not expect scientists to come out with a catalogued "happiness-inventory," there may very well be invariant necessary conditions in institutional patterns productive of happiness, that is, appropriate social relations whose violation entails a human failure to secure real happiness.

§ Invariant Tasks of Every Society

COMMON HUMAN SOCIAL PROBLEMS. The third avenue in our search for trans-cultural evaluation is to look for invariant tasks that every society undertakes in one way or another. What are the common human social problems that every society has to face? There is the problem of control over nature to provide food and shelter and ward off natural dangers. There is the need for some mode of organizing work with its apportioning of roles, and distributing goods. Again, a society has the task of ensuring its continuity by reproduction, which entails some structuring of social relations of the sexes. Some pattern of social control is required for protection against attacks from outside and disruption from within, as well as for the performance of large whole-society enterprises. Again, there are problems of the pooling of knowledge and skill and belief, transmitting customs from generation to generation, and providing bases for emotional security. Fundamental in all of these is maintaining a standardized system of communication; one is likely to forget the role of language as a magnificent social accomplishment precisely because one tends to take it for granted. The general justification for a society's undertaking the tasks lies in the fact of human needs and in the inability of individuals to satisfy them otherwise. Without some social accomplishment of these tasks, not only would society be non-existent but individual existence itself would be improbable.[34]

---

34. There has been considerable treatment of the problem of common human bases for society in recent anthropological and sociological literature.

How Well-Structured Are These Tasks for Trans-
cultural Ethical Judgment? Now granted that every
society to some degree works on all these tasks, are they
sufficiently well-structured to provide determinate criteria
for ethical judgment across cultural lines? To what extent
do they give rise to basic human social values or aspirations?

Technical judgments of comparative success may be
made to a greater or less degree in some of these across cultural lines. Different types of material culture and technology
may be compared for their productive power, degree of
control, effectiveness in the use of labor power, flexibility
and scope in applications. In organization of labor and distribution of goods there are obvious criteria such as general
rise in the standard of living (including the wide distribution of goods) and the degree to which the mode of organization operates to encourage increased production. Ensuring
continuity of the group seems to admit of a merely numeri-
cal criterion, with some qualification for health and longevity. Less precise but probably fairly determinate criteria
for comparison of success may also be elaborated for social
control, education, and even emotional security.

Nevertheless, such technical judgments do not seem to
get to the heart of the question of bases for transcultural
evaluation of social forms, nor even to be decisive in ethical

For example, G. P. Murdock in "The Common Denominator of Cultures"
(in *The Science of Man in the World Crisis*, ed. R. Linton, New York,
Columbia University Press, 1945, pp. 123-142) surveying the variety of
items in all cultures finds the invariants to lie in categories. Clyde Kluckhohn in "Universal Categories of Culture" (in *Anthropology Today*, ed.
Kroeber, University of Chicago Press, 1953, pp. 507-523) looks for invariant points of reference stemming from the biological, psychological and
socio-situational "givens" of human life. D. F. Aberle, A. K. Cohen, A. K.
Davis, M. J. Levy, Jr. and F. X. Sutton in "The Functional Prerequisites of
a Society" (*Ethics*, LX: 100-111, 1950) offer a definition of a society,
find the existence of a society terminable under four conditions (biological
extinction or dispersion of members, apathy, the war of all against all, and
absorption into another society) and then look for those functions without
whose performance one or more of the dissolving conditions results. Aberle
in an unpublished paper, "Values and Value Systems," has also formulated
the question with reference to the nature of a system of ultimate values,
looking for the elements in a society for which such a system must contain
justification.

judgment within a single society. We cannot simply say: the greater success in each field the better. In social control, ethical decision may rest on the degree of consent and co-operation and mutual acceptance of aims rather than on efficiency in securing obedience; in education, on the content and specific social role rather than the technical competence of teaching methods;[35] in questions of population, on the whole type of life that ensues from a greater population under the given resources rather than sheer increase. In some of the social tasks, no doubt, phase rules may take the form of "the more the better." Thus it is probably true that the more control over nature the better, or the more knowledge the better. But even here, specific evaluation of the performance of the task in a society, and certainly comparative evaluation of different societies, requires that the task performance be seen in closer relation to what is happening about other tasks, how biological and psychological needs of individuals are being met, and in general what is the quality of life as a whole in the given societies.

The interrelation of the different tasks may be seen from the way in which value elements from outside each task press in and seek to invade the technical criteria of success. For example, we may measure success in economic organization and distribution by rise in the standard of living. But can we rule out such factors as degree of choice in fact offered to consumers by operations of the economy, or degree of participation and initiative encouraged in the workers under that economic organization? Again, in considering familial systems in relation to the task of perpetuating the social group, can we rule out psychological criteria about what does or does not satisfy the needs of women and chil-

---

35. For an interesting presentation of varying social roles of education, see Margaret Mead, "Our Educational Emphases in Primitive Perspective," *The Amercian Journal of Sociology*, XLVIII, 633-639 (May 1943). Mead distinguishes the situations in which education is the simple transmission of ways and skills, an active proselytizing, a way of changing status, a mechanism of change or of preserving the status quo, a political device for arousing national loyalties, and so forth. She advocates "that we devise and practice a system of education which sets the future free" (p. 639).

dren? Or even the economic relations of the familial group?[36]

In spite of such problems which point the way to a more integrated mode of evaluation, focussing on tasks of a society is not without significance in comparative socio-cultural evaluation. It enables us to make negative judgments in the limiting case where a task is being completely or grossly neglected: for example, where a valuable skill like canoe-building is lost in a few generations because it is over-encumbered with rituals;[37] or where in the midst of near-starvation food is destroyed to keep up prices, as happened in the depression days of the 1930's.

IDEALS GENERATED BY TASKS. We are also able to see on a wider comparative basis how certain tasks have given rise to ideals that become part of the pattern of morality and social striving. For example, the pressure of material needs to control the physical world has in the present world found expression in the major goal of industrialization. This is now regarded almost everywhere as a good. Even the occasional exceptions seem to reflect rather special circumstances—as Gandhi's home industry served to fight dependence on British textile interests—or else to represent a romantic escapism from the attendant evils of modern industrialism. The goodness of greater human control over nature thus appears established as a phase rule, although its actual selection in a given case may depend on how far concomitant evils can be avoided or minimized.

On the other hand, the task of social control has become enmeshed in a conflict of ideals. One path follows the aim of more and more efficient control, with more and more rigid obedience; this is found in the aristocratic-Platonic tradition of rule by an élite, and recently, in various authoritarian state forms. The opposite ideal embodies the aim of less and less need for control itself. Thus in most political ideals of the modern world we find a major agreement in

---

36. E.g., Kluckhohn says: "Having several wives makes economic sense among herders, not among hunters." (*Mirror for Man*, p. 41.)

37. W. H. R. Rivers, *Psychology and Ethnology* (New York, Harcourt Brace, 1926), pp. 190-210 ("The Disappearance of Useful Arts").

expressing this aspiration. Liberalism traditionally takes all political coercion to be an evil, although necessary, and judges that government best which governs least. Communism adds to its instrumental dictatorship an ultimate ideal of the withering away of the state, replaced by cooperative economic administration. Comparable elements are found in Guild Socialism and, of course, in Anarchism with its outright rejection of state power. Democratic theory in its various forms stresses the will of the people as expressed in government, so that governing is to be less and less coercive control and more and more deliberative regulation. (Analogous tendencies could be traced in morality, itself regarded as a mechanism of social control.)

These examples show that to some extent the value aspects of the unavoidable tasks of a society can be studied directly (at least in modern ethics) in terms of the ideals found in various moral patterns. Thus, in addition to industrialization and minimization of coercion, there are ideals of justice expressing modes of organizing labor and distributing goods. Various ideals of the group and the community seek to cope on the theoretical side with the task of conceiving mankind as a continuous entity. The systematic ideals of science express the pooling of knowledge. And various religious ideals compete with secular ideals of mental health to carry out the social task of providing emotional security. Consideration of all these ideals in their cultural dimensions gains in scope and precision when they are seen as expressing, in part at least, the efforts of societies to perform their unavoidable tasks.

TASK ANALYSIS HELPFUL BUT RARELY SUFFICIENTLY DETERMINATE. We may conclude that the third path by which we sought bases for transcultural evaluation is helpful but incomplete. It gives us limits by which we can assess gross failure in cultural and social forms, and so adds some determinateness to the types of judgments made on the basis of biological and psychological needs. In some areas it provides stable ideals and may increasingly give us better measures of comparison. In others it presents major conflicts of ideals.

This is also helpful; it poses the issues and indicates the kind of evidence required, which may be available in other fields.[38] But on the whole, it marks out bounds, delimits problems, and yields general criteria rather than more determinate bases for comparative judgment or concrete decision. It can only become more specific as more determinate relationships are discovered among the various tasks in their concrete cultural performance. And this brings us to the fourth path.

§ Ethical Role of Sociological Determinants and Functions

THE PROBLEM. In what way and how far can the knowledge of specific sociological determinants and functions provide a basis for trans-cultural judgment issuing in either phase-rules or specific evaluations of the situation within a single culture? We have seen how knowledge of biological and psychological determinants and functions advanced the enterprise of evaluation, and there is no reason why this should not also be so in the social sciences. The theory of determinants is, however, especially unsettled in social science. It merges rapidly with the whole problem of method and the kinds of inquiry the material makes possible; and this in turn is bound up with general hypotheses about the structure of man and the world found in the various philosophies of history. The theory of functions is likewise in an unsettled state but has recently taken more modest shape. Functionalism as the slogan of an anthropological and sociological school has receded, and instead we find more careful empirical analysis of different types of relationships, and more explicit use of specific models.[39]

---

38. See, for example, the psychological discussion of authoritarianism, pp. 181 ff. above.

39. Cf. Robert H. Merton, "Manifest and Latent Functions" in his *Social Theory and Social Structure* (Glencoe, Illinois, The Free Press, 1949); A. L. Kroeber, "Structure, Function, and Pattern in Biology and Anthropology," in his *The Nature of Culture* (Chicago, University of Chi-

In the case of both determinants and functions, however, it is recognized that specific causal and functional judgments can be made and that, in spite of differences in analysis, many of them may be regarded as correct. This is sufficient for our present purposes. What I should therefore like to do is to see what lessons the perspective of the social sciences in this area may furnish on the question of rendering ethical judgment more determinate.

SOME SIMPLIFIED ILLUSTRATIONS. Perhaps we may best begin with several simplified illustrations. Let us suppose that in one society, a primitive community, we find a highly developed irrigation system associated with a political pattern of strict authority and a moral pattern placing a high value on obedience. In a second society we find independent farmers on good soil, with very little political centralization and a "Jeffersonian" individualist ethics. In the third there is a temporary "state" during the period of the buffalo hunt, in which for a limited time officials have the power of life and death; at other times there is a "relaxed" political pattern. In the fourth there is an agricultural community with a Demeter-Persephone religion: that is, a goddess is carried off to the underworld for half a year and returns for the other half, and without her return there would be no crop. And suppose further that morality is tied to the rules for ensuring the return of the goddess.

Now the tendency in the discussion of determinants is to look for the relation of the different "parts" of the culture —in these illustrations, economics, politics, religion, morality. Let us yield to the tendency for the moment and see what kinds of relations, logically distinct in type, one may look for.[40] Let us call them collectively *dependence relations.*

TYPES OF DEPENDENCE RELATIONS. One type of possible

cago Press, 1952); S. F. Nadel, *Foundations of Social Anthropology* (Glencoe, Illinois, The Free Press, 1951), Ch. 13; A. McBeath, *Experiments in Living* (London, Macmillan, 1952), pp. 87-95.

40. A general analysis of this question is to be found in my "Context and Content in the Theory of Ideas," in *Philosophy for the Future,* edited by Sellars, Farber and McGill (New York, Macmillan, 1949).

relation is a *causal* one, embodying hypotheses about the history of the particular culture. For example, there might be evidence that the first society had previously been without an authoritarian politics and morals, and that these had come in after the society adopted an agricultural pattern involving large-scale irrigation. This involves a causal hypothesis that the requirements of large-scale irrigation brought on or produced a ready acceptance of the centralized authority and a morality to support it. Of course, such a hypothesis would require detailed exploration of conditions under which these consequences would in fact ensue.[41] In the second society we can readily conceive of two different causal hypotheses. One has a people with a political and moral individualism come into a given area and therefore appropriate good land individually. The other has them come in with some other mode of organization, but because the land taken over is productive, they become less dependent on that organization and gradually develop individual land ownership and individualistic politics and ethics. These causal hypotheses cannot both be true of the same people or event; the issue between them is a purely factual one, requiring historical evidence.

A second type of possible relation is *functional*. Whatever the causal origin, one cultural form may be geared to the service of another in the actual living process of the people concerned. In the third society can there be much doubt that the political pattern of absolute authority is geared to the necessities of the buffalo hunt and that its maintenance in a people that relax it the moment the hunt is over in some sense "depends" on the needs of the hunt and the habits of the buffalo of travelling in large herds and the importance

---

41. Cf. "If the Pueblos represent a waterwork society in miniature, then we should look for certain authoritative forms of civil and magic leadership, for institutionalized discipline, and a social and ceremonial organization." (Karl A. Wittfogel and Esther S. Goldfrank, "Some Aspects of Pueblo Mythology and Society," *Journal of American Folklore* (January–March 1943, p. 20). See also reference to Tanala and Betsileo below, p. 241.

of accumulating a food supply for a long period in a short time? This is objectively the relation even if the people themselves should tie their temporarily changed behavior to a religious account and give it a mythological rationale. This religious pattern would then obviously be a function of the political, in the sense of an additional sanction.

Now there are no doubt serious issues of logical analysis involved in the meaning of this functional relationship. It is a kind of complicated continuous present causality in which where x is geared to y, some properties of x—at least intensity or role, if not form—may be expected to change when y changes. It is not a functional relation in the purely mathematical sense; there are too many other factors that enter into the situation. This concept is allied rather to the sense in which conduct is broadly observable as expressing purpose, whether in individual or in social life. Thus it asserts that people in the given group or society (or a sufficiently dominating part of them) cling to the servicing conduct or institution in some significant measure because these make possible a certain kind of life. Hypotheses about such functional relations are therefore at bottom proposed descriptions of *valuation patterns* of the group or society, indicating what has the status of basic aims and what a more instrumental role. Such hypotheses clearly depict a different type of relation from that of historical origin.

A third type of possible relation concerns the *content* of a cultural form. In the agricultural society with the Demeter-Persephone type of religion one could not understand the actual meaning of the religion, of its rituals and symbols, without relation to the economic-productive processes of that society. This is a closer relationship than simply a functional dependence or a causal resultant, though it does not imply that the religion may not at the same time have functions and meaning on other levels as well. It is in this sense that one may look for the economic content of political patterns or the political content of moral patterns or the moral content of religious patterns. It is also such close relations that tempt one to question at times the validity of parcelling out a culture into institutions or areas and then formulating the issue of determinants as one of their relation!

IMPACT OF SUCH RELATIONS ON ETHICAL JUDGMENT. If we think of these three relations—causal, functional, content —as *dependence relations,* we may now ask what follows for the problem of ethical relativity to the degree that dependence relations are in fact found to hold. Suppose a complete dependence in all three senses is asserted, say of politics on economics. What would this mean for ethics? In that case there could be no independent evaluation of political forms. For not only would all political events have economic causes, but all political programs would function as part of an economic program, and no political ideal would be intelligible without reference to its specific economic content. The standards of evaluation would therefore be primarily economic, and the evaluation of political forms would be as determinate as the evaluation of economic forms permitted. Similarly, if political ideals were seen in their meaning, occurrence, advocacy, etc., to be a function of a determinate theological picture, or a determinate psychological picture instead of an economic picture, like consequences would follow for ethics. The evaluation of political ideals—including related moral standards for the individual—would be referable to theological value standards, or to criteria of a healthy state of the self. For example, if the deity is pictured as authoritarian, as some religions paint it, and the injunction holds to model social relationships in obedience to divine will, then democracy wherever it occurs would represent a kind of rebellion; and so in fact the theory of the divine right of kings construed it. Similarly, if Plato's specific psychological picture is correct, in which reason is the natural leader in the soul, and political constitutions reflect the dominant state of men's souls at a given time, then democracy everywhere must be regarded as a kind of anarchy stemming from intransigent appetites of men. To the extent that the picture—economic, theological or psychological—is established as correct, indeterminacy in ethical judgment is correspondingly minimized. It is, of course, to be noted that this is a part-whole evaluation in the sense that it discovers and evaluates the congruence of the cultural form with the basic valuational structure in which it is found to be inextricably intertwined. The only remaining question of evaluation would be that of evaluating the standards of the fundamental structure itself—a large task, but a more circumscribed one.

The state of the social sciences at the present time does not establish such tight relationships. It does, however, show us partial dependencies in different senses in a variety of directions. These too will serve to guide ethical evaluation and diminish ethical indeterminacy. Thus if a political ideal like democracy is found to have a relatively independent career with changing functional affiliations and widely dispersed content, dimensions of evaluation are still provided by tracing these threads at a given time and place. For example, it may be seen that the political form has its functional role in men's demand for personal dignity, and its workings can be evaluated accordingly. Or it is held in part instrumentally with the hope of extending more widely the fruits of material progress, and this becomes a standard for estimating successful operation. Or it has as part of its content a set of concrete liberties, and their maintenance becomes a measure of its worth. Even where such dependence relations are fragmentary, they furnish guides for the ethical effort to evaluate by appraising a cultural or institutional form in terms of its wider relations and effects, by how it "works out in practice." Thus the search for dependence relations remains a methodological principle in evaluating in particular contexts even where no trans-cultural or trans-societal generalizations have been established.

SOME CAUSAL AND FUNCTIONAL RELATIONS OF MORALITY. One dependence relation that seems to be pretty well established for ethics by the social sciences concerns the broader contours of the moral character of a society. On the whole, these appear usually to be shaped into conformity with the needs and objective possibilities of an existent set of natural resources and productive processes. Such factors set the stage: they determine what is available, what is scarce, what is individually mastered, what is carried on by social cooperation, where there is a high population density or small scattered groups. One is not surprised then to find, for example, that an Eskimo society in which every man is early on his own for most purposes of securing a livelihood, had an individualistic moral pattern, without close integration; nor that societies at the heart of which lies some major cooperative task have a more integrated morality, even though such relations are never simple or regular. It is also a sociological commonplace by this time that many of men's habits stem from their occupation. Veblen has perhaps been most sensitive to the connection between habits in production and intellectual habits; these in turn provide the content of intellectual virtues such as how one is to face nature, appropriate ways to make decisions, and so forth.[42]

It is usually, however, the way in which production is organized and distribution of goods carried out which bears a greater relation to moral patterns. This seems to be a lesson of history. Perhaps it is because such questions concern more directly the relation of man to man, rather than simply man to nature. In western history, for example, the shift in dominant moral pattern is not a function simply of the contrast of landed and commercial-manufacturing economy. It also rests on the way in which the former appears historically as a self-subsistent family or manorial economy, the latter as a money or exchange economy of a growingly urban type. Such a shift had occurred on a small scale in the slave society of classical Greece, and its moral correlates were keenly felt as a contrast. Aristophanes lampoons the disparity between rural morality and the slick urbanite in jests that have meaning even in the age of television. And Aristotle contrasts the "natural" uses of money with the "unnatural" process of accumulation by charging interest in words that echo throughout later struggles of the landed-

42. See, for example, his analysis of the belief in luck, in *The Theory of the Leisure Class* (New York, Modern Library, 1934), Ch. XI.

feudal and the manufacturing-commercial moralities. Much
as historians may vary in interpretation of the shift in western Europe from the feudal to the bourgeois world, the
linked contrast of morality and economy remains. The
morality of the landed feudalism is hierarchical, a procession from lord to serf, each with his duty in his place.
With slavery no longer existent as a dominant institution,
the commercial-manufacturing morality develops universal
scope. It is an individualism authorizing the unlimited self-
aggrandisement of every man in competition with every
other, and making success (measured in monetary terms)
the final criterion. Social scientists and historians like Max
Weber and R. H. Tawney have traced in detail the shift in
virtues and goals from the ordered hierarchy of duties in
Catholicism to the glorification in Calvinism of thrift, diligence, sobriety, frugality, and made amply clear how this
was geared to economic success.[43]

These are broader dependence or functional relationships. Within a narrower time-span, the moral pattern may
be sensitive to all sorts of winds, both major conflicts that
go on in a society and the pressure of the current atmosphere. This is a ready lesson from our own times where
year to year details are more easily discernible. Thus, as
labor unions grew in strength, matching the sprawling concentration of business organizations, moral components of
cooperative social security and mutual responsibility made
inroads on the virtues of individual success and individual
responsibility for failure. The morality of laissez-faire individualism was thus weakened, and a variety of trends sprang
up alongside—the moral patterns of a more social liberalism,
of socialism, of various types of "corporate" ideologies.

Such causal and functional relations of morality laid bare
by the social sciences expand the tasks of ethical evaluation
and give them a more determinate structure. For the lesson
is reinforced that evaluation of the specific virtue-pattern of

---

43. Max Weber, *The Protestant Ethic and the Spirit of Capitalism,*
trans. by Talcott Parsons (New York, Scribner's, 1930); R. H. Tawney,
Religion and the Rise of Capitalism (New York, Harcourt Brace, 1926).

a society has as a basic component an exhibition of its relation to fundamental economic patterns in a society. Now since the latter embody fundamental goal-patterns, the result is an added dimension of evaluation. One does not merely continue to ask how far economic pursuits conform to moral precepts. One adds the inquiry not how far moral precepts conform to economic practices, but how well the morality is geared to furthering the underlying aims of men in economic life.

Similar questions can be raised concerning the relation of morality to other phases of human life—to law and politics, social structure, religion, intellectual movements, to sheer custom in all sorts of detail. And it is very obvious that in moral decision there are important interrelations. Take for example the kinds of moral answers one finds in a complex modern society to an issue like divorce. Reasoning about right and wrong in this will bring in everything from legal grounds, religious prohibitions, desirability of uniformity in various states of the Union, Freudian psychology, the right to happiness, democratic ideals of equality of the sexes, economic care of children, and so on. And it is clear, in turn, that a further relationship may be found in the various strands.

Now it is true that on the philosophical level there has been considerable study of relations. The moral aspects of specific religions and of religion in general, the religious premises of specific moralities, have been carefully explored. The ethical assumptions of political theory played a prominent role in ancient and mediaeval times when political theory was an applied section of ethics; less after the secularization of political theory by Machiavelli and Hobbes. But such studies have on the whole not been worked out in the terms and with the aims and equipment of the social sciences, although they furnish data for such investigations. For they aimed at uncovering conceptual relations in the subject-matter, not dependence relations between existent interacting parts of a culture in the actual lives of men. A central difficulty in formulating investigations in this way has been the lack of a precise mode of identifying morality itself as a definite part of a culture,[44] a difficulty which naturally carries over to the study of relations of morality to other phases of cultural life. In short, we may conclude that the major tasks of the discovery of dependence relations for morality as a part of culture still lie ahead, and it is too early to judge how far these will carry us in rendering ethical judgment more determinate.

THE RELATIVIST REJOINDER TO ANALYSIS OF DEPENDENCE RELATIONS. The ethical relativist, considering this brief presentation of some illustrative relations between moral pattern and other "parts" of cultural and socio-historical process, may be tempted to rejoinder by undercutting the scientific enterprise we have been describing and by arguing that at most it postpones the issue. To find the dependence relationships is simply to apply the criteria of one field to measure another. Suppose we could evaluate moral attitudes in the culture by political attitudes or by economic goals. This might introduce as a general mode of evaluating moral patterns, the inquiry how they are geared to performing the jobs of the various fields. But the value of the political attitudes or economic goals would not be settled thereby. Such modes of evaluation may even be accused of a conservative potential in holding uncritically to the dominant goals of the given society.

This criticism is, of course, salutory in its warning that indeterminacy in ethical judgment is not completely removed by analysis of dependence relations. Sometimes this analysis even increases the momentousness of the choice for individuals or societies by showing that particular and apparently isolated decisions are deeply rooted in whole-life patterns and may have wide consequences.

Acculturation problems and assimilation situations, group or national attempts at "modernization," "critical points" in social change, tend to pose such issues in acute form. Immi-

44. For a brief treatment of this problem and the way in which it has been approached, see my "Some Relations of Philosophy and Anthropology," *American Anthropologist* 55, 5 Part I (December 1953), esp. pp. 652-655.

grants to a new country where a different way of life is already dominant, are often faced with the task of reconciling what they regard as valuable in the old ways with what is implied by desired pursuits in the new. Comparable issues are faced in cultures living only in tenuous isolation from a dominant culture.[45] In several areas of American Indian life there is a growing self-consciousness about the rejection of the dominant American life which is regarded as aggressive and selfish by contrast with Indian emphases on generosity, sharing and anti-authoritarian equalitarianism.[46] But some integration with the dominant economic life is unavoidable, whether it be through production for a wider market, employment relationships, or inducements from the wider environment to the younger generation. Similarly, collective farm communities in Israel, on a high level of self-consciousness, find themselves enmeshed in the whole complex of political and external relations.

Large-scale value reorientations are a serious issue at critical points in social change all over the world today. The change may be necessary and even desired because of its productive advantages. To recognize in such a context that various cultural forms bear a functional relation to technological and economic bases does not itself resolve the problems. But it poses them more thoroughly.[47] Such issues

45. Cf. *Acculturation in Seven American Indian Tribes,* ed. Ralph Linton (New York, Appleton-Century, 1940). Evon Z. Vogt's *Navaho Veterans* (Cambridge, Peabody Museum of Harvard University, XLI, 1, 1951) studies the tensions in individual lives of veterans returning from World War II to Navaho communities. The values of Hopi life are brought out in sharp contrast with modern ways by Laura Thompson's *Culture in Crisis, A Study of the Hopi Indians* (New York, Harper, 1950); cf. the critique of this analysis by Robert A. Manners in *The American Anthropologist,* 54, 1 (January-March 1952), pp. 127-134.

46. An extremely interesting case is that of the Mesquakie in Iowa, who bought their own land as a community home and maintain their own ways with a strong sense that they are different and their own, in spite of constant individual economic relations with the surrounding culture. See the series of articles "We are the Mesquakie nation" by Fred Gearing (with the kind assistance of the Mesquakie People) in the *Tama (Iowa) News-Herald* 1953. I am indebted to Sol Tax for calling my attention to this situation.

47. For one phase of the problem, see above, p. 223, note 31.

have arisen at various points in the history of man, sometimes of comparable magnitude, as in the shift from feudalism, sometimes on a small scale.[48]

The relativist's rejoinder thus shows that even if analysis of dependence relations find in *every* society some major constellation of economic-social-political goals to which moral patterns are in fact geared, there remains the question of evaluating this pattern of fundamental valuation in the given society. Or again, if we find a conflict of fundamental patterns of valuation, as in a sharply split society, there is the question of mode of adjudication. And even what is functionally dependent may be adjudged so valuable that an otherwise desirable basic change may be rejected because of its deleterious effect on the social relationships.

Fundamental as these points are, they do not establish a definite indeterminacy. No doubt there will always be large components of sheer contingency in human life, and all sorts of areas in which there will be alternative *acceptable* ways. The conclusion is rather that analysis of dependence relations, important as it is in lessening indeterminacy, cannot stand alone, that it has to be supplemented by evaluation in terms of biological, psychological and social needs, as well as historical evaluation of fundamental goals and configurations. But the requirement of supplementation is no ground for overlooking the degree to which indeterminacy is diminished and dimensions of evaluation added by tracing functions. The issue, it should be recalled, is not whether abolition of all indeterminacy is possible, but whether it can be sufficiently reduced to provide standards for major choices of direction.

-----

48. A good illustration on a small scale is the change from dry rice to wet rice cultivation in the Tanala and Betsileo societies in Madagascar, described by Abram Kardiner and Ralph Linton, in *The Individual and His Society* (New York, Columbia University Press, 1939), reprinted in Newcomb and Hartley, *Readings in Social Psychology* (New York, Holt, 1947), pp. 46-55. It is interesting to note Linton's statement that "some of the Tanala tribes took over the wet rice method and abandoned it because of the serious incompatibilities it created in the social structure" (*ibid.*, p. 51).

CONCLUSIONS. We have seen thus far that there are several clear avenues along which invariant elements may be sought by the social sciences to act as a basis for trans-cultural evaluation. There may very well be invariant values of a highly general character. The use of established biological and psychological knowledge about human needs and processes has overturned the kind of ethical relativity that was based on a conception of complete human plasticity. Cultural forms are therefore capable of being estimated for their satisfaction of human needs, their interference with human processes, their contributions to human happiness. Further bases of evaluation are provided in part by seeing how well cultures perform the tasks that every society engages in. These become more specific as knowledge of social determinants and functions grows. What has already been learned shows that it is possible to evaluate a moral pattern in part by the way it fits into or helps advance basic productive and organizational aims of men in society. One more direction in minimizing indeterminacy requires exploration. It may be that the gaps in socio-cultural evaluation can be filled by historical examination to see whether there are persistent fundamental valuation patterns over a long-time span. This was the path taken by the evolutionary theorists in ideals of progress. It is worth looking to see whether such approaches are helpful, and what they may reveal. This will be done in the next chapter.

§ Is Social Science "Neutral" on Value Questions?

THE PROBLEM. Before we go on to the quest for the contributions of the historical perspective, there is one final problem raised by what has been said already. Does it add up, it will be asked, to a view that the social scientist makes value judgments and is not neutral or objective? Does he do more than describe people's valuations and himself respond as an individual, not as scientist? As Lowie once said in an aside: "The modern scientific procedure is to refrain from *all* subjective pronouncements. . . . The anthropologist as an individual cannot but respond to alien manifestations in accordance with his national and individual norms; as a scientist, however, he merely registers cannibalism or infanticide, understands, and if possible explains such customs."[49]

The debate on the social role of social scientists, whether they may venture value judgments, whether they may as such engage in social action, has been a continuous and prolonged one.[50] On the whole, the positions taken in the controversy seem to be, theoretically at least, consequences of prior assumptions and definitions about the nature and meaning of "social science," "value," "fact," and so forth.[51] We shall not therefore analyze the question afresh, but indicate by example what consequnces follow from the preceding discussion in this book.

CAN SOCIAL SCIENTISTS AS SUCH RECOMMEND ELIMINATION OF SLUMS? On the analysis we have offered, such propositions as "Slums are an evil" (where "slums" is analyzable as a descriptive term, that is, as a housing area lacking certain objective qualifications) are scientific propositions. They have a status in the social sciences analogous to psychological judgments that phobias or obsessions are evil. Evidence can be offered in both social and psychological areas. It is true that a definition of "good" and "evil" is required, but definitions are also required for "slums," "phobias," "obsessions."[52]

---

49. R. H. Lowie, *The History of Ethnological Theory* (New York, Farrar and Rinehart, 1937), p. 25.

50. For a recent symposium on the question, see *Journal of Social Issues*, Vol. VI, No. 4 (1950). R. Lepley, *Verifiability of Value* (New York, Columbia University Press, 1944) is concerned throughout with the parallelism of method in scientific judgment and value judgment. See also, Max Weber, *The Methodology of the Social Sciences*, trans. and ed. by Edward Shils and Henry Finch (Glencoe, Illinois, The Free Press, 1949), pp. 1-112; and Felix Kaufman, *Methodology of the Social Sciences* (London, etc., Oxford University Press, 1944), esp. pp. 131 ff.

51. Cf. the discussion of the relation of value and fact above, pp. 74-76.

52. On the question of definition, see above, pp. 76-79.

Does this mean that the social scientist is recommending the elimination of slums? Is he "as scientist" engaging in "social action"? Certainly in so far as he finds slums evil he is recommending that their elimination be on the agenda of human effort, if possible. He is doing this as scientist, not just as citizen. As citizen he might not have been certain whether slums were evil; he might have had the misapprehension that they encourage people to exercise their initiative in order to try to get out of them by rising in the world. He might even have thought that crowded living develops cooperative habits which separate rooms for each member of the family do not. The social scientist enlightens him by showing the actual consequences in terms of incidence of frustration and misery, delinquency, crime, etc. He quotes the statistics, the survey results, and his proper conclusion is not, "If you do not like these things, then you should regard slums as an evil," but simply "Slums are an evil."

PHASE RULES POSSIBLE.   Confusion sometimes arises because it is not realized that the judgment so offered is a phase rule, not a categorical "ought" summing up all the values in the situation."[53] For it may be the case, some would argue, that the causes which produce slums are also the sole support of pervasive goods which would crumble if slums, evil though they are, were removed by government intervention. We are, I take it, familiar with the argument that government housing support means New Dealism, which means Socialism, which means Totalitarianism, which means slavery! (We are concerned here with the logic, not the facts.) Now even if our sociologist believes all this, he can still state as a scientific proposition that slums are evil.

SCIENTIFIC COOPERATION REQUIRED.   Suppose, however, he knows little or nothing about economics, and goes to the economist, who assures him that slums can be abolished by a public housing program without any losses except the anticipated profits of a limited group of landlords over and

---

53. See above, pp. 46 ff.

beyond what they would get in reasonable compensation on condemnation proceedings for their properties. The sociologist and economist, having convinced one another, now want to know if it is "right" to deprive the landlords in this fashion of their anticipated profits. Without this assurance they are unable to assert the categorical "ought," although the phase rule "Slums are evil" stays as a scientific proposition. After wavering between the moral philosopher and the political scientist they go to the latter. He either examines their evidence or accepts their professional judgments, and assures them that the American constitution on the best available evidence allows such a housing program with these consequences if the people through their legislative organs want it. It is not only legal, but to use state power in such a way is "sound public policy" in a democracy committed to furthering the public welfare.

How does the political scientist back up this statement? He points to the fundamental goals of a democracy, to its equalitarian assumptions, to the increasing use of state power for public welfare in areas where widespread evils otherwise result, and his conclusion is, "It's up to the people." Since, however, we have so far asserted no basis for ethical validation of political democracy, our three scientists go on with the qualification, and they assert as a scientific proposition: "If democracy is a good, and if the people want it, slums ought to be abolished." Note they have progressed from "Slums are evil" to this more mandatory form.

They can, of course, go still farther. By invoking the social psychologist and his careful polls they may be able to drop the second hypothetical element, and say simply: "If democracy is a good, slums ought to be abolished." Or, even if the people do not at a given time vote for it, our scientists may rely on the proposition that people ought to want what is good for them. They might even decide to bridge the gap by public education. This they would carry on, not as plain citizens, but as scientists conscious of the obligation to proclaim the truth. That this is a scientific obligation, could, I think be scientifically established.

Perhaps we have gone far enough. Note that the doubtful issues are not whether social scientists can make scientific judgments of good and evil, but whether they can go on to make categorical judgments of what ought to be done. And this depends on whether greater determinacy can be reached in fundamental ethical judgments, such as the value of democracy. But in any case, the shibboleths about scientists not being able to make ethical judgments—except perhaps in so far as it indicates that teams of scientists from different fields would have to cooperate to reach particular policy decisions—represent uncritical inferences from an arbitrary ethical relativity.

I am tempted to add a postscript. Our social scientist wakes up one morning to find a letter from a legal firm informing him that he has inherited a block of slum houses. Does his judgment as a scientist change? If he is a consistent scientist it does not. "Slums are evil," he will still maintain. But, if he yields to temptation, he may say, "Still, they are now to my financial interest, and so I will cease my advocacy." He is still an honest scientist, yielding to temptation. If he yields his scientific integrity, he will invoke some of the rationalizations he has ˙penetrated with the aid of his fellow-scientists. This is the moral equivalent of accepting a bribe from special interests to conduct an intellectual lobby on their behalf.

VIII. Historical Perspectives

As a field of inquiry, history looks at events in both the broadest and the most detailed perspectives. We can look as through a telescope and focus on the whole process from the earliest that we know about man to the very present. Or we can focus microscopically on a particular period or even a particular episode, and watch the events centering about an individual or group.

To be fully understood, men's values have to be set in a full historical context. To find out whether men have fundamental patterns of valuation that can act as definite standards we have to look at the concrete paths of life that man has fashioned during his long career on our globe. To see how problems become structured for ethical determination we have to look into the intimate relations of the movement of events in a given period. The present chapter is concerned with the implication of such inquiries for the issues of ethical indeterminacy.

§ The Telescopic View

POSSIBLE HISTORICAL PATTERNS OF HUMAN GOAL-STRIVING. What kind of patterns of goals may we look for in the stream of history? Several types are possible and quite compatible with one another. There may be goals that have always been *consciously* sought from the earliest beginnings of man to the present day. There may be goals that have always been sought but in various disguises. There may, again, be goal-patterns showing an emerging developmental picture: they are pursued when it first becomes possible to do so, and the intensity of the pursuit increases with the extent of the acquaintance. And perhaps other goals pursued before these emerged can now be seen as their rudimentary predecessors. There are also goals which arise at a later point without clear relations to past striving; but once they have arisen, they open up new vistas, and occupy a central place. In all these types we have also to ask whether the universal or emerging goal occupied a central or a peripheral place: how deep were its roots, how wide its scope, how extensively was it intertwined with other, more varying human aims that played important parts in the value economy at different points. To the extent that a telescopic historical picture can show that there are and have been definite goals falling into any of these patterns, or comparable others that might be worked out, it provides bases for a greater determinateness in ethical judgment. For if there is anything which men have always sought, sought through thick and thin, sought in forms that were now open and now disguised, sought more eagerly the more it became known, it cannot be without some claim to rank as a good. And if its claims are not set aside, if it is not shown to be *merely* an instrument to something else, or a disguise for something else, or an inevitable compromise or sour grapes refuge, then it may even come to serve as a basic standard in ethical judgment.

Such a path of inquiry is in line with the sense of "good" outlined above[1] and employed in the several investigations of the scientific perspectives. In history especially, it is important to avoid misunderstandings of it, since the charge is often made that to draw a thread through history is to equate value with what has chanced to happen. This is to confuse our analytic quest with a simple behaviorist map-

---

1. P. 79.

ping of what men have increasingly *done*. For example, if
it were shown in some theological perspective that history
is the story of men's actual increasing entanglement in sin,
it would not follow from our account that sin is to serve as
a central standard in ethical judgment. The theological accounts themselves show the way out: they go behind man's
actions to the growing sense of his despair, the hectic pace
of his sinfulness, and thus argue his increasing sense of
alienation from the good and his inability to find what he
really seeks in the conduct that enmeshes him. So, too, the
secular historian is not bound to stop with men's acts on
the stage of history. The qualities of human experience and
the relations of act to aim and aspiration also constitute
part of the historical scene although there may be theoretical keys required for their interpretation.

LEADING CANDIDATES.   Such an inquiry has difficulties
that at times appear insuperable and often lead to despair.
Historical materials are incomplete and historical interpretation is extraordinarily variable in its guiding principles of
selection and general theories of explanation. For example,
in dealing with focal points of historical change, when it
comes to assessing the dominant strivings of men, some will
concentrate on men's pursuit of power, some on the struggle for liberty, some on material dissatisfactions. Some may
interpret man's nature operative in history in terms of a
preconceived plan certified on non-empirical grounds, and
picture the growing victory of moral perfection or the idea
of freedom.

What verification has history to offer for these varying
hypotheses concerning perennial or emerging aspiration patterns? The hypotheses themselves come from many sources,
not just from history. Power hypotheses come from political
theory and even from metaphysics, as well as from historical
description. Moral perfection hypotheses often stem from
religious sources, and Hegelian concepts of the march of
freedom from metaphysical theories of reason. Libertarian
hypotheses come more directly from historical description
of struggles and trends, coupled with assumptions about

human nature. Materialist hypotheses come similarly from historical description of trends, coupled with assumptions about causality in human affairs. All such hypotheses find expression in world-historical portraits and are depicted explicitly in the various philosophies of history. But we shall try here to detach them from their birth and line them up as candidates for the position of fundamental human aspirations in the historical scene.

It should be noted also that these candidates are not necessarily in competition for a limited place. This depends on their content. While power and liberty hypotheses seem on the face of it to be incompatible, liberty and the control of material forces could readily coexist and even be related as basic strivings. Again, there is no a priori assumption that there must be constant historical strivings, nor that any which there may be necessarily involve *reduction* of the rest to them. In the kind of inquiry with which we are concerned all judgments of centrality or importance for a given set of aspirations would themselves have to be given a specific and verifiable historical meaning.

On the whole, insofar as I am able to judge, while ideals and aspirations and strivings often appear in historical writings, systematic specialized research into these phases, with well-defined coordinates, is not far advanced. At most we must rest content at this stage with highly general results. It is worth, however, posing the problem and type of inquiry by example. Let us take as illustrations, *the struggle for power, the growth of moral perfection, the spread of liberty, the development of material control over the natural world.*

§ The Struggle for Power

WIDESPREAD OCCURRENCE OF POWER PHENOMENA. That power over others has been a widespread historical goal is clear from the long history of wars. History is often written as if peace were, in the old quip, the interval between one war and the next. National strength for making war has long been a criterion for judging questions of economic practice, political organization, educational objectives, and moral qualities. The Roman identification of "virtus" or manliness with the courage of the soldier as prime virtue is a fair mark of a great deal of human history. When we add to that the struggle for power within states, and for wealth as power among individuals, it is not surprising to find the partisan of power standing forth so often as the realist who pictures human aspirations as they are, not as dreamers would have them be.

ROLE OF POWER CONCEPT IN THE THEORY OF MAN. The concept of power has had a correspondingly great role in the theory of man. We noted at the outset its place as a major strand in the support of an individualistic ethical relativity, especially as it appears in the Machiavellian tradition.[2] Nietzsche's great philosophical influence added to this strand. His conception of the will to power sought to base it on biological and psychological as well as metaphysical foundations. His whole attack on the Hebraic-Christian tradition rested on the formulation that any non-power ideal is the expression of weakness seeking what power it can in its own devious ways.

In contemporary theory, political history and political science have been most prone to develop power themes. Bertrand Russell, for example, treats power as an impulse and finds it basic in every man: "Every man would like to be God, if it were possible; some few find it difficult to admit the impossibility."[3] He attempts to prove that power is the fundamental concept in social science as energy is in physics, tracing its forms and organization, and the modes by which it might be tamed. Similarly, ideals are often analyzed from the point of view of the power purposes they serve, that is, how they operate in keeping the masses under

---

2. See above, p. 24. For a recent review of the tradition accepting the fundamental power premises, see James Burnham, *The Machiavellians* (New York, John Day, 1943).

3. Bertrand Russell, *Power, A New Social Analysis* (New York, Norton, 1938), p. 9.

domination, or in stirring them up in the interests of an emerging power group.[4] It is not uncommon now to find political theory defined as the study of power, giving this concept the place of fundamental category in the field.[5]

CRITIQUE OF POWER CONCEPTIONS. Granted the occurrence of many of the phenomena—with some allowance for exaggeration—their significance still requires interpretation. Are they rooted in fundamental and pervasive valuations or are they secondary phenomena under special conditions? The crux of the question lies in the causal analysis of the struggle for power.

Insofar as this question is one for psychology, we have suggested that contemporary psychological theory rejects power-drives as the unique basis for human nature.[6] But other psychological bases may be used in an attempt to shore up the belief that life is fundamentally a struggle for power. It will be said that power is necessary to satisfy the mass of human desires which are limitless if given their way; striving for luxury drives us to take a slice of our neighbor's land, and him to take a slice of ours. Or it will be said that we can never be secure in what we have unless we buttress it with power after power. The first is Plato, getting ready to build the ideal state in his *Republic;* the second is Hobbes, laying the foundations for sovereignty in his *Leviathan.* Historically, and with reference to peoples rather than individuals, both seem to have a large measure of truth. But in these formulations power has an instrumental value. It is not idealized. And as the consequences of war become overwhelmingly evil in the modern world, other instrumentalities to satisfaction of desire and to security may be sought to replace the struggle for power. Plato's own account suggests the path that he overlooked. He couples

---

4. Burnham, *op. cit.*, takes such an approach to ideals. See also, James Marshall, *Swords and Symbols* (New York, etc., Oxford University Press, 1939).

5. The leading philosophical analysis along these lines is in Harold D. Lasswell and Abraham Kaplan, *Power and Society, A Framework for Political Inquiry* (New York, Yale University Press, 1950).

6. See above, p. 172.

scarcity with desire for luxury as the cause for inevitable war. This dilemma has confronted man in various ways through the ages. The ancient and mediaeval tradition, assuming scarcity as the natural background of life, stressed repression of desire, while the modern tradition has tended rather to glorify strength. But the contemporary world may have the realistic hope of removing scarcity itself,[7] thus rendering war an unnecessary instrumentality and even causing desire to lose its insatiable character. The very existence of such a prospect is enough to demote the struggle for power from its place as a measure and standard for social evaluation in the relations of men.

Such considerations lead us to suspect that the prominence of the power category in the theory of man may represent a relatively temporary (although prolonged) special set of conditions in the history of mankind, rather than a profound insight into their basic valuational patterns. Even from a short-range explanatory point of view it is doubtful whether a somewhat indiscriminate power hunger or a power maintenance drive is sufficient to furnish a unifying value account for group political behavior. Investigation will always lead beyond the power struggle to the actual goals, interests and needs of the persons or groups concerned, even in the cases where power itself has come to have a goal status in the consciousness of the men themselves.

A good test case of this principle is furnished by political ideals. The hypothesis here suggested is that a power interpretation of ideals misses their significance precisely because it thinks of their power-supporting relations as primary. A more realistic interpretation looks not merely at the way ideals function in relation to rulers or would-be rulers, but what they mean in terms of expectations, hopes, fears, demands of the people who, in the power approach, are reckoned merely as obeying or rebelling. An extreme illustration of the results of this difference in interpretation is

---

7. See below, pp. 319 ff.

seen in Burnham's treatment of the ideal of freedom from want, which he dismisses as meaningless in terms of real politics,[8] thus ignoring its tremendous strength in terms of people's basic needs for food and shelter.

The general hypothesis suggested by our critique is that power phenomena where they occur are derivative or secondary and therefore that both in psychology, historical analysis and even from the point of view of fruitfulness in political theory, power conceptions should not be given a primary place in the fundamental valuational structure of man.

It should be remembered, of course, that even if power proved to have a central place in analysis of the type indicated, other considerations of evaluation would not be ruled out as a consequence. For example, if Lord Acton's famous remark "Power tends to corrupt and absolute power corrupts absolutely" is a correct historical generalization, then this lesson would be relevant to the evaluation of even a fundamental power drive. It would impose upon men the ethical problem of what Russell called the taming of power in order to safeguard human values.

§ The Growth of Moral Perfection

SOME FORMULATIONS. Whereas the struggle for power has been, historically, a usual conscious aim, moral perfection has been consciously sought by only a small group. Their formulations have usually been theological. History is seen as the battle of the spirit with the flesh. If most men did not appreciate the full scope of the battle that was going on in their own souls, it is perhaps because they were yielding too much to worldly aims. But the battle was nonetheless real; if only men would see themselves clearly, they would know that they were striving for righteousness of spirit.

---

8. James Burnham, *The Machiavellians*, pp. 26, 117. Burnham dismisses the ideal with the quip: "Men are wanting beings; they are freed from want only by death."

Immanuel Kant's political essays towards the end of his life began to see the whole of history as a movement towards the realization of man's rational-moral nature. That is Nature's purpose, the hidden plan which the course of events is carrying out, although the battles that achieve it are prompted by selfish desires and fought in terms of worldly goals. We have here in Kant, in embryo form, the idea that Hegel developed as the *cunning of Reason* by which men's actions outrun their limited purposes to achieve the divine plan. Kant seems almost to gloat as he watches the national need to be strong for war ensure freedom of trade and industry, and thereby the consolidation of civil liberty and even religious liberty. The outcome he anticipates is perpetual peace in a federation of nations.[9]

Since Kant's time the belief that a growing moral ideal characterizes the historical development of man has been common to many idealist philosophies. That of T. H. Green, fusing Kantian and Hegelian elements, is the most notable example.[10]

NOT A PRIMARY HISTORICAL GOAL. In such conceptions the moral ideal as a goal tends either to be guaranteed on the basis of some behind-the-scenes knowledge or to be grasped by some special philosophical insight. On the face of actual history it stands out as a by-product. This is especially clear in the Kantian scheme. Only the philosophers recognize what is happening, and may perhaps push Nature's process on a little. Kant does not offer the teleological view as a scientific generalization from the study of history. In post-Kantian idealist philosophy, the special interpretation of the direction of history is likewise *added* to history on philosophical or supra-scientific grounds. It does not appeal primarily to historical justification.

---

9. See his "The Natural Principle of the Political Order, considered in connection with the Idea of a Universal Cosmo-Political History" and his "Perpetual Peace" (in *Kant's Principles of Politics*, trans. by W. Hastie, New York, Scribner's, 1891).

10. T. H. Green, *Prolegomena to Ethics* (Oxford, Clarendon Press, 1883).

It is worth noting also that the concept of a moral ideal undergoes varying interpretation, sometimes embracing as its content the whole of what would satisfy the self of man. It ceases in such cases to have the primary meaning of perfection of character. For example, McBeath, to whose empirical analyses of moralities we have referred in the search for invariants,[11] treats the moral ideal as the good for man, the system of goods (the ends of desire) that will satisfy the whole nature of man, and maintains that moral goodness or goodness of character is not aimed at directly.[12] By separating moral goodness in this way and identifying it with loyalty to the recognized ideal, he is led to conclude that moral goodness has been a constant throughout human society. This hypothesis guided his investigation of particular cultures. Thus there is no development or progress in moral goodness, but only in the actual content of the ideal—increasing enlightenment concerning the extent of the moral community, the conception of personality, the meaning of cooperation, wider provision for the needs of human nature, and so forth.[13]

Even with respect to moral character it may, however, be true purely as a historical generalization, that the course of events is operating in one direction. Mankind may be led with increasing probability to the development of a particular kind of character or ideal of character, just as different national goals at different times have led particular peoples to hammer out and standardize a specific character mold. But in tracing such a development the by now familiar cautions must be observed. The picture must be well-grounded empirically, not postulated on rationalistic grounds prior to history as the path that history must take. The causal analysis should show the convergence of forces producing the character resultant, and it should distinguish the emergence of the character from the emergence of a positive value-attitude towards the character even when

---

11. See above, p. 217.
12. McBeath, *op. cit.*, pp. 60-61.
13. *Ibid.*, p. 435.

historically these come together. And evaluatively it should leave room for passing judgment on the character in terms of the problems to which it provides an answer. Accordingly, primary concern will tend to fall on analysis of the supporting historical forces and goals, pointing beyond the character trend itself.

§ Liberty as Historical Aspiration

CENTRAL CONCEPT IN LIBERALISM.   The spread of liberty as the universal goal was the favorite of 19th century liberalism. History is read as the struggle between Liberty and Authority, whether the actual participants in the past were directly conscious of it or not. In the earlier history of mankind, when all sorts of restraints on the individual are necessary for group cohesion to ensure survival, liberty is conceived chiefly as protection against the tyranny of rulers. Only at a much later point is it articulated as the right of an individual to self-expression in whatever form he wishes, subject to public safeguard of others' similar rights. But once it is so conceived, then it is easily seen as the force underlying earlier struggles. And it is justified in turn not merely by the existence of human desires but by the way such liberty works to achieve greater social welfare and human happiness.

Such is the picture one derives from J. S. Mill's *Liberty,* if one looks to the historical introduction and occasional historical remarks. And a comparable picture can be found in a great part of the writings of liberalism in almost any field of inquiry. Hobhouse, looking back on the history of liberalism, asks whether it is a constructive or only a destructive principle: "Is it of permanent significance? Does it express some vital truth of social life as such, or is it a temporary phenomenon called forth by the special circumstances of Western Europe, and is its work already so far complete that it can be content to hand on the torch to a newer and more constructive principle, retiring for its own part from the race, or perchance seeking more backward lands for missionary work?"[14] And he finds an answer in terms of the idea of growth and the self-directing power of personality—"the opening of the door to the appeal of reason, of imagination, of social feeling."[15]

Similarly, in recent liberalism, John Dewey gives a central place in his social outlook to organized intelligence releasing in increasingly positive fashion the energies of men.[16] At times Dewey formulates his outlook in a historical manner by speaking of organized intelligence expressed in technology as the genuinely active force in social transformations.[17]

Is Liberty a Historical Aspiration or an Independently Certifiable Ideal? The question before us is not the general issue of the desirability of liberty, but how far it can be seen to be a definite historical aspiration of mankind. On this there has been marked disagreement. Even in the 19th century, some thinkers began more and more to see the historical process as the advance of the striving for equality, with liberty as something precious to be defended against inroads.[18] Others, moving towards socialism, began more and more to see the ideal of liberty in the form advanced as a reflection of a laissez-faire capitalism with its desire for unrestricted, individual enterprise.[19]

The 20th century has seen the 19th century concept of liberty challenged widely, in theory as well as in application. The rise of fascist outlooks led many to assert that the mass of the people want to be ordered rather than to

14. L. T. Hobhouse, *Liberalism* (New York, Holt, 1911), p. 19.

15. *Ibid.*, p. 122. For a historical analysis of the relation of liberty to other components in the liberal outlook, see G. DeRuggiero, *The History of European Liberalism,* trans. R. G. Collingwood (London, Oxford U. Press, 1927).

16. Cf. John Dewey, *Liberalism and Social Action* (New York, Putnam's, 1935).

17. *Ibid.*, pp. 74 ff.

18. An interesting study of such tendencies to contrast liberty and equality will be found in B. E. Lippincott, *Victorian Critics of Democracy* (Minneapolis, University of Minnesota Press, 1938).

19. For a recent formulation of this view, see Harold Laski, *The Rise of Liberalism* (New York, Harper, 1936).

be free. The spread of communism led many to the view
that the mass of the people are governed more by material
needs than an appreciation of liberty. A socially oriented
liberalism need not, however, abandon prematurely the hypothesis that liberty does correspond in some sense to a
deep-seated human aspiration. It must, however, in order
to render the hypothesis meaningful, provide a clear conception of the liberty that it reckons with. It is hardly likely
that such a conception will coincide with the particular form
that liberty took in 19th century society, especially in its
denial of a positive role of government in promoting social
welfare. It is more likely to be construed in terms of a
general picture of intellectual and personal growth and social cooperation as Dewey envisages it, with rich opportunities for responsible individual initiative.

Now it would be quite meaningful, once such a conception is clearly formulated, to propose the hypothesis that
mankind has always in some degree had some aspiration
to liberty, that wherever it has been kindled under otherwise "undistorted" conditions it has invariably burst into
flame, that men who have truly experienced it cannot cease
to value it, that where it is rejected it has not been properly
tasted or experienced, that where more elementary survival needs are met pressures for the extension of liberty will
unavoidably increase.

Whether such a hypothesis is correct is, so far as I can
see, by no means determined as yet. It involves a fuller
history than has yet been written of the struggles for liberty in all ages and the awakening that liberty has brought
at various times in human history. But it is equally clear
that the arguments so far urged against this hypothesis
oversimplify its theory and neglect its implied conditions.

§ Development of Material Control
over Natural World

DESCRIPTIVE THREAD CLEAR, BUT ETHICAL INTERPRETA-
TION DIFFICULT.   In purely descriptive terms the clearest

historical thread runs through the growth of man's material control over the natural world of which he is a part. The historical "telescope" view shows this directly. There is the continual growth of man's material means of coping with nature, sometimes slow but occasionally with revolutionary leaps as, no doubt, in the discovery of fire, of farming, of metals, and recently with break-neck speed as in the development of atomic power.

The material hypothesis is that in some sense or other there has been a perennial human effort to develop man's productive control over nature. But what does this represent in ethical terms? How basic has this motif been in the actual life of man? Has it had unrestricted primary scope as a dominant value? Or has it been, while constituting a running historical thread, at all times subservient to other values? For example, has it been merely a question of material means to provide for maintenance of life so that higher spiritual values might take over? Or has it been incidental to a variety of other strivings at different times, e.g., to biological needs in earlier times and to men's power desires over other men at later times? Or—to take a still different major possibility—has it served first as a means and then grown into an end? All these are historical possibilities. They cannot be disposed of by impressionistic speculation or introspection. The answer can come only as the result of historical investigation and analysis.

Several types of studies—anthropological, historical and philosophical—tend to show that the goal of developing material control over nature has a central rather than a peripheral or incidental role in human striving. First, it is commonly recognized that apparently unrelated areas in life often have a material role to play. This at least establishes it as one constant dimension in human striving. Some have also found a substantial material base and some definite material content for ideals of the spirit. A third direction has been to attempt to find governing causal relations between men's material life and the remaining phases of their life.

Do Apparently Non-Material Areas of Life Have a Material Role? Religion is the best example of an area of life which seems on the face of it to be furthest from material needs and efforts. Yet a scientific study of its functioning shows religion is often intensely practical in its aims. In primitive societies it frequently has a direct concern with the preservation of the food supply, with animal and human fertility, with warding off disease, and so on. These roles are shed when secular knowledge takes over such functions. Everywhere it has had functions of moral control, that is, shaping of specific attitudes to act as girders in the cultural pattern. These are usually, when looked at not in terms of their generalized psychological character but in terms of their specific historical content, geared to the major social institutions and problems of the period.[20] Most pervasive, perhaps, have been the emotional functions of religion—providing security of feeling in the face of man's ultimate fateful situation. This point, much stressed today in discussions of religion, is still a practical one. But it does indicate—and this is a lesson which the growth of knowledge likewise establishes—that no sharp line can be drawn between man's aim to control his environment and his need for gaining control over himself, that the practical extends beyond the physical and biological into the psychological area of human life.

Material Bases for Ideals of the Spirit. A comparable outcome attends historical study of the ideals of the spirit, in fact of ideals generally. It shows that they are, when fully spelled out, attempted organizations of men's energies to satisfy basic needs in the light of existent conditions. Their spiritual quality is not belied by evidence of their material historical bases, but is enhanced thereby, for the major human ideals are thus furnished stable roots. The best example is the history of the ideal of knowledge itself.

---

20. See, for example, H. E. Fosdick, *Great Voices of the Reformation* (New York, Random House, 1952), esp. his general introduction, and the introduction to the selections from Luther.

In the ancient philosophies of Plato and Aristotle, knowledge was made pure and turned into the ideal object of contemplation. Such contemplation is more akin to vision than to scientific investigation or discovery. The whole dominant ancient attitude to the purity of speculation, it has often been noted, reflects the separation of the mind from the material world. This conception of knowledge as purely an intellectual goal was challenged at many points in the development of modern thought—from Francis Bacon's stress on knowledge as power to Dewey's instrumentalism resting on a biological-evolutionary conception of the intellect as a tool for solving problems. The application of the physical sciences in remodelling our natural environment has undermined the older contemplative attitude and the development of atomic power has shown most dramatically the inter-relation of remote mathematical equations and practical accomplishment. Developments in the philosophy of science itself have overcome the sharp separation hitherto existent between rationalism and empiricism as philosophies. Hence, contemporary assessments tend more frequently to assert the linkage of theory and practice.

Knowledge loses nothing of its height as an ideal nor its intensity in human consciousness by being rooted in man's drive for control over nature. In fact, as he becomes conscious of his aim and its bases, the growth of his aspiration becomes more clearly marked. What began as the simple increase of knowledge to control pressing aspects of the natural world (e.g., the discovery of fire) grows into the striving for the widest systematic knowledge and control. It goes on into the psychological and social sciences as well, into history and into an understanding of man's own aims or self-knowledge. The outcome then is no longer a simple effort for material control over the natural world. It becomes rather an *aspiration of mankind to gain increasing knowledge and control over nature and over itself.*

HAS MATERIAL LIFE A GOVERNING CAUSAL ROLE? In addition to these approaches that give an important position to man's striving for practical control over nature and over

himself, the further attempt has been made to find a governing causal relation between material life and the remaining
phases of human life. Such an attempt need not involve
"reducing" these other phases to the material. But in such
a view the categories of material culture are given a central
explanatory role in the panorama of cultural unity and historical change.

Actually, historical description has tended to fall almost
naturally into categories from the domain of material culture, technological processes and organization of production. Thus the archaeological-historian uses the demarcation
into technological ages. V. Gordon Childe describes the way
in which basic early inventions carried with them different
modes of getting a livelihood and consequently different possibilities of life.[21] In the Old Stone Age men were limited
to gathering and hunting, without control over their source
of food supply. In the New Stone Age they already had
planting and herding; the population accordingly increased.
The Bronze Age had specialized skills and traders; the discovery of iron as more readily available and cheaper than
bronze made possible more widespread clearing of forests
and more extensive development of agriculture. And anthropological description regularly sets its scene in terms of the
way in which a society secures its livelihood in specific
environmental conditions. It also explores both the dynamic
and the limiting role of technological-ecological factors.

Such an approach is not limited to the treatment of
early societies. It is a widely accepted part of sociological
and historical theory and technique of analysis in dealing
with all periods of human life. Thus it is not unusual for a
writer on rural and urban sociology to regard occupation
as a "first variable" that carries with it a host of other cultural differences.[22] And it is a commonplace to find economic
forces given a major place in accounts of historical devel-

21. V. Gordon Childe, *Man Makes Himself* (London, Watts and Co.,
1936. Reprinted by Mentor Books.)

22. See, for example, P. Sorokin and C. C. Zimmerman, *Principles of
Rural-Urban Sociology* (New York, Holt, 1929), pp. 56-57.

opment, although not all historians would go as far as Charles Beard when he complains that economic accounts of the industrial revolution "do not trace to their utmost ramifications the influences of technology upon the very woof and warp of civilization, including poetry and art."[23]

GOVERNING CAUSAL RELATIONS IN THE MARXIAN THEORY OF HISTORY. A large-scale attempt to establish the primacy of material conditions in social relations and human culture is found in the Marxian theory of history. This regards productive relations based on a definite stage of development of material productive forces as the foundation on which social, political, religious, aesthetic and intellectual life rise as a superstructure. Marx stresses the conditioned character of consciousness in the superstructure and the way in which it responds to changes in economic conditions of production.[24]

Marx is maintaining in effect that men are in some sense concerned with the drive to extend and maintain their control of the natural world in all their superstructural social forms. The latter may appear in consciousness more vividly, but the former constitutes the underlying reality. This is a complicated relationship: in part the economic base causes the form of consciousness, in part consciousness serves a functional role for economic forces, while in part the superstructure embodies a relatively independent set of values that can be set up against the economic forces, but cannot win in the long run, presumably because they have no comparably independent source of energy.

The ethical outcome of Marx's approach to history is that various superstructural forms would be evaluated by their role in relation to the underlying economic processes. With reference to a given society in a definite period this

---

23. Charles Beard, is his introduction to J. B. Bury, *The Idea of Progress* (New York, Macmillan, 1932), xxi. It is worth noting, however, that Beard's description of technology shortly after already included a philosophy of nature and "great constellations of ideas."

24. Cf. Emile Burns, ed., *Handbook of Marxism* (New York, International Publishers, 1935), pp. 371-372.

appears to yield an ethical relativity. For the Marxian theory sees a conflict of classes with opposing economic aims, and therefore opposing bases for evaluating the remainder of the cultural forms—political, legal, religious, intellectual, aesthetic. This relativity disappears, however, once the broader historical picture is added. For class aims are themselves evaluated in the given society and period by the degree to which they release the development of productive power; hence they are described as "progressive" or "reactionary" in the given context.

CORE-CONSTELLATION OF PRODUCTIVE POWER, KNOWLEDGE AND INSIGHT. Such extreme theses of the primacy of material conditions and objectives in human life of course raise serious problems of logical analysis and historical validity. But the ethical inferences that may be drawn concerning historical material goals do not depend on these extreme hypotheses as premises. They follow from the minimal points on which there seems to be wide agreement— that there is an historical effort on the part of men to develop their control over the material world and themselves, that other areas of life, including the life of the spirit and the intellect, turn out to have some practical bearing on this quest, and that this quest does have some causal weight in the patterning of social life whatever the precise limits of its influence and the independent weight of other aspects or phases of life.

This very minimal common consensus concerning a historically discoverable valuational pattern in itself provides a powerful base for evaluation. It embodies a three-fold criterion—*productive power, knowledge,* and *self-consciousness* or *insight* and these are felt as ends as well as powerful means. The fact that they are not the only ends of men should not lead us to minimize their strength as a standard. They constitute a powerful core constellation, by which whole modes of life or social organization can be evaluated. Of course, there has to be a specific temporal reference, since the operations of a system at one time may be advancing production and knowledge and insight, and thwarting it at another. Similarly a particular institution or belief (e.g., a religious dogma or political platform) may be examined for its help to or hindrance of scientific advance. And various beliefs may themselves be evaluated as "rationalizations" in terms of their role and relation to the realities of the situation, if they exhibit lack of insight and self-consciousness.

§ Standards Discoverable in Historians' Value Criteria

APPARENT DIVERSITY OF HISTORIANS' CRITERIA. So far we have looked for invariant directions of striving and aspiration as history reveals them. We have not yet, however, looked to see whether any agreed-upon standards of evaluating the life of a people are to be found in the judgments of historians. The initial negative that one would be tempted to give is too hasty an answer.

It is true that what strikes the comparative observer is the variety of evaluative criteria in judging different civilizations. One historian will look to buildings, wonders in sanitation, and technological skills. Another will concentrate on military prowess. A third equates high civilization with trade and economic prosperity. A fourth looks to political cohesiveness, a strong sense of participation, and active local government. A fifth counts the works of art or estimates the quality of the cultural products. What one brands as corruption—the extension of individual pleasures—another sees as progress.

Such criteria are fragmentary, and call for attempts at unification. But the patterns of unification likewise show a variety. One seeks unity in economic terms, another in political, another in moral, and so on. In this fashion the variety of initial criteria may be reproduced on a theoretical level. This should not lead to a conclusion of ultimate relativity in the philosophy of history, any more than it did in the theory of culture. It would simply mean that if there is a solution, it lies in further development of the philosophy of history itself, and a more careful study of the kind of evidence which would justify one philosophy of history as against another.

We need not, however, be led into this avenue of inquiry. For we are not seeking criteria of a "high civilization" but ethical criteria for assessing the actual life of a people, whatever be the "height" of their civilization. In many respects this is a social analogue of the inquiry into happiness in the case of an individual. That also, we saw, was a difficult question, one only beginning to be capable of empirical study in the light of the growth of psychological knowledge; and the dominant approach as yet was negative, that is, in terms of identifying the conditions under which it did not flourish.[25] The same situation may very well hold in the case of the life of a people.

HISTORICAL USE OF CONCEPT OF "DECADENCE." There is a vast literature in different branches of historical interpretation in which attention is fastened on the phenomenon of "decline" or "decadence" of a people and its works. There are distinctions between "Golden Ages" and "Silver Ages," between the spring and summer of a culture and its fall and winter. There is almost universal historical veneration of some limited periods, such as Athens in the mid-5th century B.C. and Shakesperean England. The concepts in which such judgments are made are usually far from precise. They are often metaphorical, drawn from the health and prime of the organic body. Sometimes, when one surveys what has been meant, for example, by "the decline of Rome" it seems as if the only point on which there is agreement is that Rome was conquered by the barbarians. But in almost all cases where "the civilization" of a people declines, the people live on and have descendants. The decline therefore means in the first place that some of the works the prior generations accomplished are not accomplished by the succeeding generations. But it is not merely the products to which the judgment refers. Vague as it may be, it has

---

25. See above, pp. 184-186.

reference ultimately to some actual change in quality in the life of the people.

CREATIVE POWER. The hypothesis I am suggesting is that there is one criterion almost universally recognized in such historical estimates of the life of a people, and that difficult as it is to identify and vague as it is to conceptualize, we may term it *creative power*. We see it best in extreme contrasts by comparison with its absence. The Athenians as described in Pericles' famous funeral speech[26] have it; the Spartans do not. It is characterized by a kind of irrepressible energy, widespread individual initiative and readiness to assume responsibility, a sense of collective purpose, a society in which people know what they are working towards, rich cultural expression in a wide variety of fields, spontaneity coupled with a deliberative rationality, absence of general fear and anxiety no matter how pressing realistic dangers become. Historical judgments of particular functioning institutions or works of art or literature may not always locate this quality accurately, but I suspect that is what they are looking for. And it is possible that an energetic winnowing of the evaluations of historians, especially of the arts, would show it. Meanwhile, then, as a hypothesis, *creative power* is added to the evaluational criteria that emerged from our consideration of historical directions of striving and aspiration.

A special significance attaches to this criterion in the fact that it implies some measure of widespread individualism.[27] The criteria of productive power and knowledge are quite compatible with a division of labor in a society whereby a few have the knowledge and central skills, whereas the rest carry out instructions. (The whole society may, of course, derive the benefits; we are not considering distribution patterns at this point.) But creative power, to be characteristic of a society, rests on widespread cultivation of the inner resources of individual men. It is possible

---

26. Thucydides, Bk. II, ch. 36-43.

27. I am indebted to George Boas for underscoring the distinction that follows in this particular context.

that insight, to a marked degree, shares this character. And to some extent the progress of knowledge itself depends on a wide critical spirit and appreciation. It follows, since the creative power here described is not a biological trait, that a study of the conditions that foster it may add further criteria for more determinate ethical judgment.

§ The Microscopic View:
The Problem of Specific Relevance

NEED FOR HISTORICAL CONCEPTS IN JUDGING RELEVANCE IN PARTICULAR ETHICAL SITUATIONS. We now put aside the telescope and move towards the microscope. What conceptual tools has historical science fashioned for specific valuation in the particular—beyond the application of the universal criteria so far suggested—so that ethical judgments for a given epoch or a given people or a given individual may be more determinate? Logically, the problem here may be expressed as one of finding a mode of judging *relevance*. We may approach the particular armed with a host of phase-rules or other generalizations about what is good, what is right, what men aspire to. But these go in different directions at times, and may actually conflict. At best, values bewilder by their rich variety. How can we decide which to invoke as appropriate to a given time and place?

Such a formulation suggests that the historical events of a given time and place constitute a kind of problem matrix to which moral rules and proposals may be referred and in terms of which they may be in some fashion evaluated. An approach of this sort to evaluation in particular cases is useful only on two assumptions. The first is that goal-striving and ethical demands generally do not arise at random in human life, but are an expression in significant degree of problem-situations in which men find themselves. The second requirement is that problem-situations be themselves to some degree well-structured. If they are not, indeterminacy will remain in the problem itself and be reflected in a conflict of goals or a shifting from goal to goal.

For as we saw in originally posing the problem of indeterminacy, its chief sources lay in field instability and field
complexity.[28]

The first assumption is, in my opinion, sufficiently established by the sciences of man. Goals do have natural bases
in problem-situations. The functional approach we discussed
above in biology, in psychology, and in the field of society
and culture embodies this view. It is not yet certain how far
the second requirement is met. The structure of the problem-
situation is not always clear. We do find psychological
needs operative in individuals, and social needs and tasks
operative in society, as well as perennial aspirations in successive generations of men. But whether these structure
particular problem-situations sufficiently to minimize indeterminacy in major ethical judgments in particular contexts
seems to be a question largely for historical evidence. Is
there enough evidence of sufficient structuring to enable us
to speak of *the central problems* of a people, of mankind in
a given epoch? Are these scientific concepts, or merely persuasive and argumentative notions?

USE OF RELEVANCE CONCEPTS IN PHILOSOPHIES OF HIS-
TORY.   Philosophies of history that accept a deterministic
pattern readily employ such concepts. The determinism may
be theological, idealistic or materialistic. For example, in
the Old Testament, once God has decided to liberate the
Hebrew people in Egypt and bring them to the promised
land, there is no doubt about *the central problems* which
determine ethical judgment. In this framework which involves the complete acceptance of God's will, everything is
seen as right which brings this to a successful resolution,
including the death of the first-born of the Egyptians. And
any complaints on the part of the Hebrews that they might
have suffered less by staying in Egypt is simply weakness
in response to their sufferings in the desert. There might,
of course, have been profound differences of policy on how
best to get out of Egypt, or what course to pursue in the

---

28. See above, pp. 97 ff.

desert, but the terms of the central problem are tightly structured.

In a Hegelian idealist philosophy of history the plan of world-history is set, but each period has its own share of the tasks to accomplish. Heroic men are the instrumentality: "Such individuals had no consciousness of the general Idea they were unfolding, while prosecuting those aims of theirs; on the contrary, they were practical, political men. But at the same time they were thinking men, who had an insight into the requirements—*what was ripe for development*. This was the very Truth for their age, for their world; the species next in order, so to speak, and which was already formed in the womb of time."[29] Hegel adds that these men, when their object is attained, fall off like empty hulls from their kernel; Alexander dies young, Caesar is murdered, Napoleon is transferred to St. Helena. But the philosopher, as Hegel sees it, can be wise only after the fact—"The owl of Minerva takes its flight only when the shades of night are gathering"[30] —and see the central tasks which determine the spirit and accomplishment and morality of the period. To this all problems are referred. Everything else Hegel regards as utopian dreaming.

In Marxian materialism, the historical driving force is the development of productive forces. To this the functioning of the superstructure, including morality, is related. Hence the pivot for moral problems lies in the existent stage of development. Marx develops this conception in a striking passage in the introduction to his *Critique of Political Economy*: "A social system never perishes before all the productive forces have developed for which it is wide enough; and new, higher productive relationships never come into being before the material conditions for their existence have been brought to maturity within the womb of the old society itself. Therefore, mankind always sets itself only such prob-

29. G. W. F. Hegel, *Philosophy of History*, trans. by J. Sibree (New York, P. F. Collier and Son, 1900), pp. 77.

30. *Hegel's Philosophy of Right*, translated by S. W. Dyde (London, George Bell and Sons, 1896), xxx.

lems as it can solve; for when we look closer we always find that the problem itself only arises when the material conditions for its solution are already present or at least in the process of coming into being."[31]

Liberalism, on the other hand, has tended in its philosophies of history to avoid such "agenda of history" concepts. It looks upon them as part of tightly knit intellectual systems produced a priori for special purposes. It opposes the theological and idealist patterns in terms of its own scientific premises; and it opposes materialist patterns in the name of empiricism and pluralism. Liberal theory thus often tends to share the ethical relativist's denial that there is a determinate historical structuring to the moral problems of an age or a people.

A scientific empirical approach cannot subscribe a priori either to a monism or a pluralism in a particular area of fact. In principle it sees the question of whether unified patterns or diverse configurations can be found in the moral problems of an age as an empirical one. And it is to some degree separable from the specific theory, idealist or materialist, which develops it. If upon scientific investigation a tight determinism turns out to be correct, then there certainly is an agenda of history which at a given age poses problems of social evaluation upon all, whether they realize it or not. If there is no simple determinism, there may still be a set of problems in each given age which in fact structures moral issues whether people realize it or not. Their connections may be looser, their underlying forces may be more diverse, with larger contingent elements. But if they exist, then their moral import is extremely significant, for they provide a more determinate principle of particular ethical judgment.

EXAMPLE FROM THE HISTORY OF ENGLAND. Let us explore the relation of ethical issues and historical problem-situations more carefully by example. Take English history

---

31. *Handbook of Marxism*, ed. Emile Burns (New York, International Publishers, 1935), pp. 372-373.

as furnishing successive problem-situations. And as an example of an ethical principle, take the development of the equalitarian idea. For brevity's sake let us limit the account to political equality, assuming its goodness as a phase rule; certainly in some contexts its moral quality stands out, although in others it may be seen chiefly as a technical device. We then have two questions to ask corresponding to the requirements noted above for an ethically well-structured historical situation. Does the actual career of political equality in English history show that striving for it developed as an expression of the historical problem-situations of the English people rather than arising arbitrarily? Secondly, in looking at these problem-situations, can we make objective judgments (for which historical evidence may at least in principle be offered) of irrelevance, peripheral relevance, spurious relevance, genuine relevance, central relevance, overwhelming relevance (and so on) of political equality to the historical problem-situation?

The first question concerns a descriptive picture and its causal analysis. It asks for the extent and scope of the ethical issue[32] over a historical period and for its causal basis. Let us look briefly first at the actual career of the principle in English history. Obviously there was no issue of political equality before England had some political structure. Again, neither King John nor the Barons nor the people of England accused one another of violating equalitarian ideals, nor was Magna Carta set down as a document for later historians to argue about its "democratic" implications. It was in the late 1640's that the issue of every man having a vote was raised by the Levellers. The Levellers lost out to those who held that a vote went only to men with property interests in the kingdom. But political rights had been raised as a moral problem, a claim of every man in virtue of the fact

---

32. Note that the reference is to the issue not the acceptance of the principle. Political equality as an ethical issue would have a wide extent even if a great part of the population saw it as the central moral wrong, the chief enemy, instead of as a good. Its scope refers to its ramifications and the degree of its entanglement with different areas of life.

that he, like everybody else, had one life to live. The debates of the Levellers with Cromwell's representatives remain a source of moral insight as well as political wisdom.[33] With the coming of the Industrial Revolution, the tremendous increase of population, and the creation of a large factory proletariat, the issue of political equality moves to the center of the scene in the consciousness of the people. A central ethical judgment in the consciousness of the age now rests on whether institutions, programs, attitudes, do or do not favor equalitarianism. This becomes the basis of moral-intellectual alignments.

Today, political equality is not raised as a central problem in England, nor is it a widespread criterion for ethical judgment—though it threatened to become one when élitist conceptions appeared on the horizon in fascist forms in the 1930's. It is still, of course, a high value, tied in with many attitudes in the English moral pattern. But its maintenance or reaffirmation does not seem to settle one way or the other a vast host of moral problems facing the English people. The scope of the principle has thus diminished today. There was a time when the demand for political equality rocked England. It no longer does so.

Now in surveying this career the historian constantly probes for underlying causal forces in the needs and alignments of the people. He may draw a social map of the demand for equality in the 17th-century English revolution, for example, and show precisely who were the people who supported the Levellers, what their religion, what their occupation, and so on. Similarly, surveying the emergence of equality demands on the center of the English historical stage in the late 18th and 19th century he may inquire to what extent they rested on a general growth of humanitarianism, to what extent on the wider productive possibilities of the new technology, to what extent on the alliance of industrial capitalism and labor in the former's bid for greater

---

33. See, for example, William Haller, ed., *The Leveller Tracts 1647-1653* (New York, Columbia University Press, 1944).

political power. Again, he will want to know why equalitarian political demands reached the center of the stage before equalitarian economic demands, and what the relation of the two has been.

In general, then, both the historian's mode of inquiry and the extent of success in such inquiries give an affirmative answer to our first question: the emergence, extent and scope of human goal-striving do express in significant measure the historical problem-situations of men. But precisely how this yields greater ethical determination carries us into the second major question. The historian's inquiry itself moves over into this area when he begins to make critical judgments of the materials revealed in the first inquiry. Some of the hopes of people at certain points, although they operate as historical forces, are judged as "misguided." Some aims are "premature." Some are "Utopian insights" which only later on become realistic policies and programs. Such judgments are extremely complex, but they may be capable of empirical certification. For example, a hope is misguided if people acting on it assume it will have certain consequences which in fact in those circumstances it could not have. A policy is premature if the necessary conditions for its understanding or acceptance or implementation are lacking though it does point towards the successful resolution of the underlying problems. For example, it may be debated whether the Levellers' struggle for political equality was premature. Those who deny it was so will say that it was a demand which would have resolved the problems to which it was addressed and England would have been that much ahead, but unfortunately it lost out through historical accidents. Those who regard it as premature will say that it expressed a fairly narrow social base, that the demands for greater economic freedom could then be met without going so far while the needs of the agricultural laborers and dispossessed peasantry (which found expression in the Diggers' demand for abolition of private landed property) were not reckoned with by the purely political equalitarian ideal, that moreover when the political equalitarian ideal became strong later on in Utilitarian and Chartist movements it had a different social complexion.

Such inquiry carries one deeper and deeper into the actual historical evidence for the structure of pressing problems of the given age. It is only as the problems become clear and as they are found to have a determinate structure that the judgment of the degree of relevance of the principle to the problems can be carried out. This will determine whether the principle takes its place in the moral economy as simply adding an increment of value or whether it is to be an overriding standard by which character-attitudes, specific goals, economic structures, educational programs can be evaluated. Now the concept of a historical problem-situation is already itself a fusion of factual and value components. It is not value-free, but this is not question-begging since the principle to be estimated for its relevance is not included. Thus the principle of political equality becomes an overriding standard when the pattern of human needs in the given period (economic advance, technological skills, health demands, necessary educational demands, psychological demands, etc.) converge on a wide scale in such a way that political equality is in fact in that situation the key which will open the door to exercise of human energies towards the achievement of many needs. Such situations have sometimes arisen, whereas at other times the structure of problems appears looser-grained. The degree of relevance of the principle then may undergo change.

A fuller study of historical relevance in ethical judgment would have to make clear the specific types of interlocking of valuational and descriptive elements in the concept of historical problem-situations. It would have to examine the way in which biological and psychological needs as well as socio-cultural and general historical values become integrated with the specific materials of a given period into a dynamic structure which rules out some possibilities, makes demands in the lives of men, shapes valuations and orients strivings. Such a study would have to be sensitive not merely to the valuational elements in the struc-

ture but also to the assumptions in critical ethical judgments
about the structure which may reflect a wider historical
generalization. For example, to say that political equality
had no relevance at a given time does not imply that it would
have been less valuable if it had somehow or other entered
the scene; it is a judgment comparable to the "iffy" proposition that certain things would have been possible in ancient
times if men had had modern technical knowledge. It probably entails the assumption that we cannot say someone
ought to have aimed at some particular thing if it was in
fact "impossible" for him to know about it. Similarly, to
say that a man or group who realized the historical "impossibility" of a particular struggle ought nevertheless to have
continued it entails a general assumption about the value
of self-sacrifice over a long historical span. Even protests
against tying ethical judgments too closely to historical
problem-situations can be seen as reflecting historical lessons about the conditions under which men tend to be
shortsighted or put on blinders.

That configurations which "over-determine" ethical judgment are to be found in the flux of history cannot be considered established for all periods, but they do seem to
characterize some periods. Yet the general hypothesis cannot be ruled out. It remains a heuristic principle upon whose
fruitfulness depends in part the degree to which ethical
judgment of particulars is rendered determinate. Once
again, it is not to be seen as an all-or-none question. Judgments of irrelevance or peripheral relevance may be possible
even where precise judgments of central relevance are more
difficult.

JUDGMENTS OF CENTRALITY IN CONTEMPORARY ENGLAND.
Take, for example, the present situation in England. Has it
the kind of integration which makes it possible to say in an
historical sense, "England's major task today is such-and-
such?" Many formulations do take this path. Some say that
the central problem is to keep a firm hold on what remains
of the empire. Others see it as raising the standard of living
of the people of England. Some pose it as a transition to

socialism. Others pose it as resisting socialism. Some see it as fighting communism, and some as bringing on communism. Some even insist that the central task is moral rearmament.

Now if the concept of central problems in a given epoch is to be of any scientific utility, the choice between these alternatives cannot be simply a matter of arbitrary will, nor wholly a choice between ultimately conflicting values. It must involve a descriptive-historical question of the actual fundamental valuation-pattern in England during the given epoch, set in a picture of existent forces, to the best of our available psychological, social and historical knowledge. In such a historical investigation some assumptions about values will be taken for granted and some values may be turned up which would require judgment outside of the context of inquiry. For example, it would probably be assumed that raising the standard of living in England is wanted by the people of England and is a good, and the inquiry would be concerning the centrality of this task. On the other hand, suppose it were shown that there is no realistic alternative in the light of costs, risk and disposition of forces in the world, to a relaxation of empire controls sooner or later. Then obviously the ideals of empire would to that extent lose historical relevance apart from whatever outside judgment might be given of their ethical quality.

The decision which of these proposed tasks has a central place, which is peripheral, which is completely irrelevant, or even whether there may be no central tasks in the structuring of English life and problems in this period is a historical one, within the limits of certain common value assumptions. It is almost as if each candidate were saying "The things you people want or value or have to have can now be maintained or achieved only by working in this general direction, or are integrally bound up with actualizing these particular values." But even where there is a supreme *value* conflict involving major cleavage of outlook, and even if it breaks out in actual revolution, this does not deny—in fact it tends to confirm—a kind of central-problem structur-

ing of the situation. For it may constitute major agreement
about what is the central *issue* to be settled, without agreement on the answer.

Whatever the historian's answer as to which issues are
central, once we have an answer there is a more determinate base for ethical judgment. If the ideals of empire are
historically irrelevant today, any ethical judgment, for example that Englishmen should turn into Spartans to keep
the empire, would have a spurious historical quality even
apart from a further ethical comparison of the lives of Spartans and the lives of Englishmen. If raising the standard of
living is a central task, then the moral quality of holding in
private ownership for hunting enjoyment land which might
be used for agricultural purposes is seriously affected—unless, of course, the issue becomes the wider moral problem
of the absoluteness of private property rights. A view of the
transition to socialism as the central task calls for abandoning many hitherto established principles of the rights of
ownership and profit, hence for a substantial revision of the
moral concepts of social justice, with a strong emphasis on
equalitarianism as the central thread. A view of the central
task as stopping socialism calls for the virtues of tradition,
patience and caution in making social changes, fostering the
individual sense of social responsibility so that it will carry
out the social welfare tasks that otherwise in a modern society are handed over to government (e.g., the spirit of
charity), and so forth.[34]

We need not go through the ethical implications of all
the hypotheses mentioned. Those in terms of fighting communism and extending it will be touched in the wider global
context in which they are usually set. That a moral rearmament thesis calls for far-reaching alteration in moral outlook
needs no elaboration.

The evaluation that results from determining what constitutes central tasks and problems, where in fact life is

---

34. Compare, for example, John Parker, *Labor Marches On* and Quintin Hogg, *The Case for Conservatism* (Penguin Books, 1947).

structured so as to present them, is of course never the complete or total evaluation. There is always the possibility that some further value will counteract its results. Situations of personal tragedy sometimes arise from the fact that the central issues of a time make irrelevant a high value to which the individual is completely devoted. In a society as a whole this is less likely, since the initial formulation of the test for centrality involved a mapping of the valuational picture. Hence the existence of such a value on a broad scale would enter into the structuring of the situation and the greater the devotion to it the more the central problem of the epoch might become finding a place for it or resisting its rejection. Opposition to tyranny and the struggle for the liberty of a people from oppression has sometimes achieved its centrality in precisely this fashion, whether the struggle in fact proved successful or ended in destruction.

GLOBAL USE OF RELEVANCE CONCEPTS. The importance of historical relevance concepts in adding a dimension of determination for ethical judgments may be illustrated further from global formulations of *the central problems of mankind in our era.* It is interesting to note how widespread in practice is the use of such a concept. It seems to recognize that whatever philosophy of history be employed, there is a meaningful descriptive-historical sense in which there can be a convergence of forces such as to thrust some problems onto the center of the stage. And it makes a tremendous difference which is regarded as the central problem. Compare, for example, current analyses of the global picture. The dominant American analysis sees the central problem in terms of a battle for the maintenance of free institutions against Communist aggression and conspiracy. The Communist position presents the central problem in terms of the effort of a declining capitalism to maintain its power by force against a spreading socialism which will abolish exploitation and lay open the way for mass progress. A prevalent approach in India sees the central problem in terms of the effort of impoverished peoples of the globe to achieve improvement in their lives through acquiring the technology, knowledge and skills of the west without falling under the domination of either America or Russia.[35] A global formulation along the same lines that had wide currency during World War II spoke in terms of the rise of the common man everywhere to independence and self-determination, seeing the Russian revolution as simply one incident in the efforts of undeveloped countries to achieve modernization and looking upon the 20th century as primarily the century of the common man.[36]

These are only a few of the different perspectives about what are the central problems of our day. Now even apart from different stresses in value that may be incorporated into the views, there is a common area of historical happenings and forces they are trying to interpret. And in their conflicting aspects they cannot all be correct. Meanwhile each becomes the basis of a different moral recommendation. Those who see conflict as inevitable call for militarized strength and hence invoke the virtues, goals and controls of the armed camp. At an extreme point, tolerance is regarded as weakness and compromise as near treason. In the third view cited, stress falls on planning for relief of impoverished areas and technological and industrial progress. Tolerance, cooperation and even readiness to compromise become the prime virtues.

SOME CONCLUSIONS. The conclusions pointed to by our extended illustrations may be summarized briefly. The first and most important is that there is a meaning to the concept of *the central problems of mankind in a given era,* or to *the central problems of a country.* The second is the hypothesis that at some times at least this concept has application, that sufficiently unified patterns are found in the valuation-existence relationships in a country or on the globe as a

---

35. This approach is to be found in the statements of Nehru in India. The extent of its prevalence in the East is well conveyed by Eleanor Roosevelt in her account of her travels in the East. See her *India and the Awakening East* (New York, Harper, 1953).

36. Henry A. Wallace's *Century of the Common Man* speech of May 8, 1942 expounded such a conception.

whole. But, in the third place, the picture as found in consciousness, no matter how widespread, must not be equated offhand with the picture as it really exists. There can be a false consciousness of the central problems as well as a true consciousness. The determination of the true one is a matter for empirical investigation, with history playing a significant role in the judgment. However, if historical knowledge itself contains an inherent indeterminacy there may remain an indeterminacy in the ethical judgments that rest upon it. Therefore how the issue of contrasting historical accounts and interpretations is to be construed is one of the crucial points upon which the extent of the possibility of diminishing indeterminacy may be seen to rest.

The problem whether historical knowledge is itself inherently "relative" to purposes, interests and biases of investigators is itself part of the wider question of the objectivity of knowledge in the light of the cultural and historical conditioning of the process by which knowledge is acquired. It is important to consider this wider question, at least in outline.

§ Is Knowledge Itself Culturally and Historically "Relative"?

THE GENERAL ISSUE.   Briefly, the central issue can be stated as follows. The diminution of indeterminacy in ethical judgment has been secured in various ways, chiefly by invoking results of science or pointing to possible solutions as a consequence of further knowledge if it could be attained. But what would be the consequences for ethical theory if knowledge itself turned out to be culturally and historically conditioned? Since the cultural and historical conditioning forces already embody values, would not the resulting more definite answers in ethics provided by scientific scrutiny simply reflect the operation of these deeply hidden values?

We have noted at several points the impact of such an

approach.[37] There was the attempt to declare reason a response to irrational forces operative in the individual. There
was the view that cultural elements pervaded even fundamental theoretical concepts, especially through the selection
operative in the culturally determined language and its influence on mode of thought. And just above, the issue was
raised whether different historical pictures of even current
events may not represent irreducible value perspectives.
Even the gradual accumulation of a body of stable knowledge in the physical sciences may be seen on an extreme
cultural relativist view as reflecting not truth increasingly
forcing itself on men but simply the stabilization throughout the world of a single cultural tradition.

IN WHAT SENSE IS SCIENCE A CULTURAL-HISTORICAL
PRODUCT? Studies in the cultural and historical relations of
science have proved very fruitful in contributing to historical and cultural knowledge. Without either mapping their
scope or estimating their hypotheses, we may make a few
distinctions that are very important to clarifying the general
issue before us.[38]

In the first place, a distinction in the cultural-historical
treatment of science may be drawn between the *causes* of
science, the *uses* of science, and the *content* of science. In
connection with the latter, a further distinction may be
drawn between the *meaning* of scientific assertions and the
truth of scientific assertions.

Now if all events have causes, and scientific assertion is
a phenomenon in nature, it too will have causes. Scientific
investigation of such causes has already shown that some
of these causes are cultural and historical. Thus the historian
traces the way in which the practical problems of a period
give rise to certain scientific efforts and assist them in reaching fruition, and so on.

Again, since science is a part of culture, the uses to which

---

37. See above, pp. 52 f., 206 f.
38. I have dealt with this problem in greater detail in my "Context
and Content in the Theory of Ideas," in *Philosophy for the Future*, ed.
Sellars, McGill and Farber (New York, Macmillan, 1949), pp. 419-452.

it is put, the purposes and goals involved, have cultural and historical components. Questions of the application of science, the suppression or encouragement of discoveries, the utilization of scientific views in political struggles, and so on, are obviously largely cultural and historical phenomena.

If we turn from causes and uses to content, it is quite clear that some scientific assertions—for example, those in the social and historical fields—have a cultural and historical content. That is, the meaning of many terms may require reference to specific cultures and specific periods.[39] For example, to understand any assertion about "freedom" in the social and historical sciences requires specific cultural-historical reference, although we have seen that this need not exclude invariants for all such references. And it may very well be the case that there are more cultural-historical references required in the analysis of scientific assertions than has customarily been recognized.

Even in the physical sciences, one might take the system of physical theory at any time and separate strands which have both a cultural-historical genesis (cause) and some cultural-historical content. Indirectly the instruments used in verification can be seen as social products and embodying some of the specific limitations of the society at the time. The linguistic-communication aspect of theoretical ideas may reflect some of the fundamental attitudes of the culture. The regulative principles may embody philosophical assumptions and follow philosophical models that have elements reflecting the culture—for example, a teleological interpretation in the earlier history of physics. And so on. Thus the state of science at any one time may be shot through in various ways with elements of content reflecting the particular cultural-historical way of seeing and doing and thinking. Even the way in which scientific assertions are tested may be found to reflect cultural assumptions of the degree of evidence warranting belief in different areas.

---

39. For the role of unique particular-culture variables, see above, pp. 207-211.

As a result, the system of scientific belief at any time is in some sense or at least in part, a cultural-historical product.

BUT SCIENTIFIC TRUTH IS NOT A CULTURAL-HISTORICAL PRODUCT. I have elaborated the range of cultural-historical penetration into science in order to set off sharply what appears to me quite definitely not to be culturally-historically dependent. And the chief of these is the most important one—the truth or falsity of the scientific assertions. It is obvious that causality and genesis do not determine truth or falsity. Similiarly, it can readily be seen that use does not determine truth or falsity; on the contrary, it is truth or falsity which will in the long run determine the scope of the use that will be successful. Where cultural and historical elements come closest to the truth problem is when they enter into the content of the scientific assertions. But here they refer to and help clarify the meaning of the assertions whose truth is to be determined. They do not determine whether what is asserted, once it is asserted and its meaning fully clarified, is true or false.

There have, of course, been attempts in the theory of knowledge to relate truth in some domains to the special perspective of the inquirer. Mannheim's sociology of knowledge adopts in part such an outlook in dealing with historical-social knowledge.[40] But it is difficult to say that such views have been made even clear enough to be seriously considered.[41] Their essence usually lies in unmasking particular theories and showing that they embodied a limited value-perspective.

Now it is this very phenomenon which seems to me to be further evidence for the objectivity of truth. As we saw,[42] the very unmasking of a bias as irrational in individual psychology or as a special group interest in social history pre-

---

40. Karl Mannheim, *Ideology and Utopia* (New York, Harcourt Brace, 1936). See especially, pp. 258 ff., 266.

41. For a detailed critique of the relativist view in historical methodology, see Maurice Mandelbaum, *The Problem of Historical Knowledge* (New York, Liveright, 1938).

42. P. 53.

supposes that criteria are possible for differentiating rational from irrational, group rationalization from group insight, and so forth. The more unmasking, that is, the more discovery of special values influencing judgment, the more the criteria become sharpened in empirical terms to distinguish the real from the masked. In short, the fact that science is in many senses a psychological product, a social and cultural product, a historical product, in no way deprives it of a trans-personal, trans-cultural, objectivity.

METAPHYSICAL CONSIDERATIONS AND PRACTICAL DIFFICULTIES. It may be said that the position here taken issues from a particular kind of metaphysics—one in which there is a belief in an objective reality which forces itself increasingly upon the attention of men and brings their wanderings eventually into line. Such a fundamental outlook does simplify the issue before us, but it is not my intention to invoke it here. On the contrary, it is possible to view such an outlook as in significant degree an extrapolation from the definite results of scientific knowledge, the development of its world picture and the stabilization of its method of inquiry.[43] In a more pragmatic vein, necessities in the biological makeup of man, in the decisiveness of action required for survival force upon man a concept of truth which cuts across cultural lines. In any case, no special metaphysics need be invoked to give scientific knowledge an objective character. The stability of physical knowledge today does not rest on either a realist or a positivist or an instrumentalist interpretation of its concepts, but is a fact cutting across all perspectives and expressing a pervasive order in the natural world. If the knowledge of man and society and history reached anything like the stability of physics we would similarly have a far greater degree of determinacy in ethical judgment irrespective of diversity in cultural interpretations.

On the other hand, practical difficulties often lie in the

---

43. Some indication of such an approach is to be found in my "Interpretation and the Selection of Categories" in *Meaning and Interpretation* (Berkeley and Los Angeles, University of California Press, 1950), pp. 57-95.

way of accumulating knowledge, whether in psychology or the social sciences or history. There is no single formula for resolving such difficulties. If they stem from insufficient investigation, inadequate concepts, interference with inquiry, and a host of similar factors, the deficiencies may sooner or later be overcome. If they stem from excessive field complexity and field instability, they may be indicative of a basic indeterminacy. But it is too early in the history of psychology, the social sciences and history to accept the second alternative. In any case, current shortcomings need not be elevated into a matter of principle. On such issues, the long-run history of physics, starting with the clash of schools in ancient Greek philosophy of nature, is both illuminating and comforting.

§ Applications to Historical Knowledge

THE SPECIFIC PROBLEM IN HISTORY. Because of the complexity of historical materials, the possibilities of diverse selection, the multiplicity of theoretical approaches, and the close relation of value assumptions to methodological procedures, the problems we have discussed in general are particularly aggravated in history. When history is rewritten in succeeding ages, much of its revision is at the same time an unmasking of what was previously regarded as "objective" history. As a consequence objectivity is itself sometimes brought into disrepute, and its maintenance as a goal is sometimes taken to be a mark of an undiscriminating dogmatism with regard to one's own assumptions.

If such an attitude is carried to an extreme, its outcome is a complete historical indeterminacy. It may be said that each fundamental perspective is correct "from its point of view." Sometimes it is added that the victorious side writes the history books, or that one can believe any of the opposing views in our day by a sufficient "will to believe," or that we are too close to events in our time to be able to estimate them accurately or that men will always distort the picture to suit their interests.

SUCH ARGUMENTS DO NOT ESTABLISH A BASIC INDETERMINACY. That men have sometimes distorted accounts to suit their interests or made mistakes under the influence of conflicting passions does not establish an inherent indeterminacy in historical knowledge. On the contrary, the lesson is the same as in the other fields we considered: the more we learn about such happenings the more we have specific tools for correcting such distortions in other cases. Even the influence of a will to believe can be studied carefully to determine the limits within which and the conditions under which hopes and fears may tip the scale.

That history is constantly rewritten reflects, as is often noted, the advance of the various special sciences which increase the store of tools for reconstructing historical knowledge, as well as the growth of hypotheses from a longer-range point of view.

That we may be too close to contemporary events for historical objectivity does not itself wholly preclude historical lessons advancing ethical determinacy. Certainly in many questions of the past, enlightenment has come long after the fact. And the lessons of men's historical aspirations which we examined in the telescopic view may be learned from the more established "neutral" materials of the past. Nevertheless, even with regard to the present, we need not accept the Hegelian dictum that insight comes only after all is over. As the importance of historical knowledge to ethical judgment is realized, the importance of better tools for its acquisition in contemporary affairs becomes a factor in the human situation. To diminish the gap in understanding contemporary events may itself become one among the central problems of an age. For example, the very inability to resolve the problem of contemporary historical knowledge and the conflict of points of view may itself make almost an absolute value out of the freedom to inquire; for only through free inquiry can the conflict of interpretations yield a fuller truth as a basis for ethical judgment.

PRACTICAL TECHNIQUES EMBODYING DIVERSITY OF INTERPRETATION NOT AN ADMISSION OF FAILURE. Every field has to work out its own paths for achieving the widest truth available at any stage of its growth. If in a particular area of history the best results so far can be obtained by giving free rein to opposing interpretive principles and producing widely differing accounts of the same events,[44] it does not follow that objectivity is surrendered. On the contrary we have here the raw materials for a more objective picture, for such a presentation may help to analyze and ultimately remove the particular and partial biases in the interpretive principles. When such "relativity" is recommended as the best way to get a more truthful picture, it has no commitment to a doctrine of ultimate inherent indeterminacy in historical knowledge.

What the future may bring in the development of historical inquiry is as yet unclear. Too many actual interferences stem from the conflicts of the modern world. But historical objectivity remains a goal, and one upon which the path of diminishing ethical indeterminacy is in some measure dependent. In any case the actual contributions of history should not be minimized. History takes its place with the biological, psychological and social sciences, as a partner in providing bases for ethical judgment.

---

44. For an interesting illustrative study of shifting interpretations of a single set of events, see Howard K. Beale, "What Historians Have Said About the Causes of the Civil War," in *Theory and Practice in Historical Study: A Report of the Committee on Historiography* (New York, Social Science Research Council, Bulletin 54, 1946), Ch. 3. Beale's conclusion is that with a desire to learn, "knowledge of history can teach this or any generation a great deal about itself and its problems." (P. 91.)

PART THREE

Towards
A Common Ethic

IX. The Theory of the Valuational Base

§ Taking Stock

TAKING STOCK OF OUR FINDINGS. We have found in surveying the perspectives of the human sciences that ethical indeterminacy could be diminished in a number of ways.

One way consisted in establishing phase rules or other *cf. P. 46* ethical generalizations. These were of a wide variety of types, dealing with general goods, virtues and character-constellations, obligations and tasks, strivings and aspirations, but also with values that might have a high degree of specificity. For example, we saw how it might be possible to decide on the goodness of life, pleasure, health, drive expression, and the evil of pain and frustration. We saw how fundamental human needs were traced and how some candidates for this position were rejected. We saw how a growing knowledge of personality development might make it possible to generalize about such character-constellations as the anti-democratic personality, the positive value of a variety of human attitudes, such as cooperativeness, and some of the ingredients in an abstract goal such as happiness. Generalizations were also seen to be possible through phenomenal description and analysis which revealed common human feelings and attitudes in varying socio-cultural forms, and a host of further possible directions in the search for invariant trans-cultural values was indicated. Judgments of invariant social evils were also recognized as possible, and to some extent judgments of invariant obligations or tasks of societies. In addition there were phase rules found in the historical domain, expressing perennial aspirations (such as productive power or knowledge), non-perennial but possibly invariant values (such as liberty) and possibly invariant criteria in judgment (such as creative power). Our inquiry showed that the possibility of establishing generalizations was a rich one, though on the whole the harvest was yet to be gathered.

A second way of diminishing indeterminacy in ethical judgment consisted in the elaboration and multiplication of theoretical tools for inquiry and analysis. In part, these enabled us to discern phase rules even where there was surface complexity and variety. But they also proved applicable to ethical judgment in particular contexts. They are not then merely tools for the discovery of generalizations but tools for applying generalizations in particular evaluation. For example, we found it possible to utilize the discoveries of contemporary psychology to analyze or discover an individual's values beyond his introspective account of them. Again the functional perspective in all the sciences furnished detailed methods of analysis. For it correlated description of striving and value experiences with knowledge of causal processes. As a result it focussed on the relation of a human trait or action or social form to the basic aims and patterns of mankind, the individual or the culture. Such analysis of function or role added a fresh dimension of ethical evaluation. Applied in the historical field it suggested a verifiable meaning for the concept of central problems of an age or people and thereby helped furnish tools for judging the relevance of phase rules to specific situations.

In addition, the simple application of logical and analytic methods proved a powerful solvent in dealing with traditional obstructions to inquiry. For example, insisting on an empirical analysis of the self in doctrines of egoism compelled their translation into a form in which scientific evidence could more readily be brought to bear on their evaluation. And in many other areas logical analysis showed the presence of psychological, socio-cultural and historical variables within the very heart of ethical formulations.

All these findings showed that no a priori bars should be set up to the contributions that the several sciences can make to rendering ethical judgment more determinate, that the contributions so far are already considerable and significant, and that once the channels for ethical-scientific cooperation are cleared, considerable further advance may be hoped for.

THE OPEN DOOR TO NEW FINDINGS AND NEW FORMULATIONS. Central in such an approach is the maintenance of an open door to new findings on the part of the sciences and new formulations on the part of ethical theory. Not one of the areas which we have explored as the basis of extending determinacy in ethical judgment can afford to close its books. It must reckon not merely with new hypotheses but with constantly changing conditions that provide fresh ground for testing its established views. Fresh evidence about human needs or psychological dynamics may have fresh ethical implications. Similarly a changing historical scene may necessitate marked shifts in the formulation of central problems.

New formulations in ethical analysis may also pose fresh inquiries. For example, when the concept of the phenomenal field is applied to ethical theory and the question is raised whether there are invariant values in the phenomenal field for a given situational meaning[1] a new set of inquiries is asked for, quite different in type from the older search for invariant acts. Similarly the historical search for perennial objects of striving is broadened considerably when we ask not merely "What have men always worked towards?" but also "What have they always valued highly after they first came to experience it?" To ground human values in some account of human nature and the human situation

---

1. See Duncker's formulation above, p. 194.

need not involve overlooking the possible development of human nature and possible changes of quality in the human situation. An intuitive realism of Hartmann's type sometimes expresses this phase more accurately when it thinks of values as analogous to existent stars that are undiscovered and sees the telescope turning on fresh areas of the sky. Or again, sometimes a mountain-climbing model is suggested. New situations raise us above older levels, and the vista that bursts in upon us is not less true because it could not be anticipated from the previous position.[2] Such formulations are not bound to one philosophical approach. The full understanding of what is discovered as a new value may come only in seeing it in terms of man's needs and problems in a fuller setting; but recognizing such a value may itself bring a refinement in the conception of these needs and problems.

New findings and new formulations in questions of ethics are not the task of science or ethics alone. Nor can there be a simple division of labor with the findings assigned to one and the formulations to the other. A constant collaboration seems required if there is to be progress in diminishing ethical indeterminacy.

NEED FOR SOME INTEGRATING CONCEPT. One fresh type of construct is suggested by the summary of our findings. The variety of phase rules, the variety of tools for value analysis, the several modes of relating the rules to the particular context of application constitute a formidable array of materials and instruments. In any actual time and place, from the point of view of decision, they can be invoked and where relevant applied. Now we may ask what the sum of the lessons of such application amounts to, what the growing area of determinacy actually is which can be used as a basis of judgment. In short, if possible, it would be desirable to have an integrative concept by means of which to set forth what unity actually exists in the lessons of inquiry and

2. This metaphor is used with telling effect in dealing with the analysis of needs, in Solomon E. Asch, *Social Psychology* (New York, Prentice-Hall, 1952), p. 342.

to formulate a standpoint for evaluation. For this purpose I should like to introduce and experimentally to explore the concept of the *valuational base.*

§ The Concept of the Valuational Base

WHAT IS A VALUATIONAL BASE? The concept of the valuational base is intended to point to the crystallization of the determinate elements we have found, and to suggest that these may play a fundamental role in evaluative processes. It is a base because it acts as a standpoint for evaluation, providing moorings to which a morality may be fastened, and sailing-charts for its general course. It is valuational because it embodies major decisions of policy supported by an interlocking structure of human knowledge and human striving. It is not simply identical with basic value premises widely held or pervasive attitudes, since even a strong and widespread value inclination may be distorting human needs and be shot through with false assumptions about the world. The valuational base may be seen rather to contain value conclusions or guiding principles embodying the fullest available knowledge about men's aspirations and conditions.

The constituents of the valuational base will clearly represent a fusion of universal and local elements. The universal are the fundamental human needs, the perennial aspirations and strivings, and the discovered high values that are deeply grounded in these needs and aspirations. All these will appear not in their bare abstracted psychological or generalized social form, but in the richly varied shapes that human development in all its dimensions has given to them. The local elements will be the central problems of the age including its central necessary conditions and special contingent elements (not necessarily values themselves) which may press heavily upon men's pursuit of values in the given age. The specific task in working out the valuational base for the contemporary world is to locate the constituents and trace the particular pattern of their linkage.

WHAT CAN BE EVALUATED FROM THE STANDPOINT OF
THE VALUATIONAL BASE? Before examining in greater detail the constituents of the valuational base it is worth looking to see the kind of use to which it may be put, that is,
what areas of human life may be brought within its evaluational scope.

It is obvious that judgments of the desirability or unde-
sirability of specific social policy can be referred to the
valuational base. Both institutions and social programs can
be evaluated by or found good to the extent to which they
provide satisfaction or expression of fundamental needs, 1
conform to perennial aspirations, function so as to solve 2
central problems by meeting central necessary conditions 3
and facing crucial contingent issues. 4

The current morality of a people can be evaluated in a
similar fashion, at least in some of its aspects. A morality
includes characteristic approved goals or goods and evils,
rules of duty and obligation, virtues and vices, and so on.
In all these, dimensions of evaluation are provided by asking
the questions: Do they express or hinder satisfaction of
fundamental needs? Do they further perennial aspirations?
Are they geared to the solution of central problems? And
so forth.

Even the ethical outlook of a people may be subject to
evaluation in terms of the contents of the valuational base.
For the ethical outlook of a group, which includes its principles or modes of justifying or criticizing its morality, involves a conception of the world and man. For example, a
moral virtue of honesty may be justified in one ethical outlook on the assumption that it is God's will and men's task
is to obey that will, in another on the assumption that
honesty is the best policy in attaining happiness that all
men strive for, in a third on the assumption that it sets a
framework for "good" human relations. The knowledge involved in a valuational base may be used to judge the correctness of the factual assumptions of the ethical outlook
as well as to evaluate its effects in the lives of its adherents.

Such a concept of the valuational base is applicable to

any particular society of men and similarly to any period in human history. Specification of the local condition is, of course, required. The standpoint from which the base is set up is contemporary knowledge—on the view we have attempted to establish that growing human knowledge does provide a standpoint for evaluation. The phrase "the valuational base for a given period" or "the valuational base for the contemporary world" indicates simply that the local elements considered are those of the given period. It is not to be confused with the value orientation of the specific society as an anthropologist would describe it.[3] In fact the value-orientation of a specific culture can itself be evaluated by reference to the valuational base.

The knowledge embodied in the valuational base is, of course, open to alteration or refinement in the growth of knowledge itself, just as some asserted scientific laws of one time are sometimes found later to have been loose or inaccurate beliefs.

Both the types of constituents of the base and their evaluative role may be seen more clearly if we consider them briefly one by one.

UNIVERSAL NEEDS. These, precisely because they are universal, can be counted on to be present and operative, taking distorted forms when their direct expression is denied.

In contemporary formulations universal needs are described in biological and psychological terms, such as survival, need for affectionate warmth, and so on. We have also seen that once we begin to use them for evaluation, they rapidly furnish more than vague invariants on an abstract level. Moral qualities in interpersonal relations, such as sympathy and cooperativeness appear to have a positive value in terms of their satisfaction of psychological needs and their relation to developmental values in the growth of a person. Thus the psychological and social sciences today

---

3. Cf. Clyde Kluckhohn and others, "Values and Value-Orientation in the Theory of Action," in Talcott Parsons and Edward A. Shils, ed., *Toward a General Theory of Action* (Cambridge, Harvard University Press, 1951).

have brought the new insight that growth rather than disciplined order by repression or a childlike dependence on superiors is the type of development which makes the individual more capable of happiness, in a sense which includes maturity and fulfillment. And this in turn provides criteria for evaluating social institutions and cultural ways. It even affects the evaluation of ethical outlook. It suggests, for example, that a formulation of ethical concepts in terms of an individual accumulating pleasures is less satisfactory than one construed in terms of gratifying modes of human relatedness.

Stable results in morality, social policy and even ethical outlook may thus be the result of following the thread of fundamental needs. If any specific values become firmly enough established in this way and have ramifications wide enough—especially through relevance to central necessary conditions and critical contingent problems—they may even be incorporated into the valuational base for contemporary society. They would then not only help estimate but even point directions for the satisfaction of fundamental human needs. Thus in the contemporary world we may use the moral criteria of deepening human sympathy and extending the sphere of human cooperativeness without tracing the connections to need satisfaction in every context of application.

PERENNIAL ASPIRATIONS AND MAJOR GOALS. Which among men's values, aspirations and goals are to find their way into the valuational base?

It is worth noting that values in human life are many, and value experience is widespread. The sum of values extends far beyond the sum of morality, ethical outlook and valuational base all put together. A full analysis of their relation would require a treatment of concepts of value, means, end, far beyond the scope of the present work. For our present purpose we do not even have to ask whether perennial aspirations such as knowledge serve as means or ends or both, or each in different contexts or periods.

Furthermore, a value does not simply find a place in the

valuational base because of its height. It has either to move
men and serve as a goal or have very wide ramifications in
their lives. It need not generate problems; it may be thoroughly intertwined with the socio-cultural structure, and
find so adequate a satisfaction as to be taken for granted.
Oxygen and food are both bodily needs of a comparable
order. The need for oxygen is usually satisfied automatically;
it has no ramifications and only a marginal value status. Under normal conditions it does not belong in the valuational
base.[4] Provision of food, on the other hand, is obviously a
goal in the valuational base; for it has a major place in the
structure of any society's activity even where there is material abundance. Similarly, the love of beauty and the desire
for friends are human values of a high order. In the modern
world the desire for friends may rise to a high status in the
valuational base because of the intensified loneliness of men
in the social structure of the contemporary large city. The
love of beauty has a perennial quality and a high value;
how deep are its roots and whether it has a secure place
in the valuational base are questions for a psychology and
sociology of appreciation and creative experience.

Lack of a place in the valuational base may not therefore
be construed as derogation of value. It may mean simply that
the value is a relatively isolated one, not sought but appreciated highly when it happens to be actualized. This question
may, however, be less important than it seems if it turns out
that most high values do express fundamental needs, since
this would give them some connection with the base.

Of the perennial aspirations and major goals some may
have an invariant human character—for example, the core-
constellation of historical aspiration discussed above. Many
culturally developed forms may express fundamental needs
and have universal application—for example, the goal of
growing into the particular language upon which communication with the group depends. Many major goals will
vary from culture to culture and still have in each a depth—

---

4. It may, however, belong to the *personal* base of an individual who
is a permanent invalid. For the personal base, see below, p. 329 f.

rooted in their relation to fundamental need satisfaction—which gives them a place in the valuational base. For example, in modern times the desire to have a home and family satisfying as it does a whole complex of fundamental needs, might be given a place in its own terms in the base.

To map this segment of the valuational base in the contemporary world is a task calling on the most refined combination of philosophical analysis and scientific description and explanation. But that all these goals and aspirations can function in an evaluative way should be fairly evident. Social conditions can be assessed for their impact on the possibility and depth of friendship. Cultural and even moral attitudes can be measured for their harmonizing or disruptive effect on home life. And even the perennial aspirations in their most general form can serve as forcible evaluation points. Consider, for example, whether it does not approach to moral condemnation of a social program, moral attitude or even ethical outlook today to say of it that it would stop scientific progress and lead men to put on blinders.

CENTRAL NECESSARY CONDITIONS. The great means of a period are the basic conditions (in their specific cultural and historical form) without which men will be unable to achieve their values, diverse as the latter may be, in the conditions of life, knowledge and civilization of that time and place. Peace, internal order, a job in an industrial society, education today, all have this character. Means of importance, because of their scope, loom large in the valuational base, often overshadowing universal but relatively isolated values which may play a comparatively smaller role. Large-scale means may acquire a value quality and become constituents of the life taken to be good.

Such central necessary conditions are not to be read off easily from the consciousness or beliefs of the people or equated with the prevailing tendencies. For example, the magical rituals of primitive Australian peoples for ensuring reproduction of food animals do not constitute realistic central necessary conditions for their satisfaction of hunger—leaving aside whether in the given culture they constitute necessary conditions for emotional security. Similarly in the modern world, while industrialization is pretty clearly a central necessary condition to be incorporated in the valuational base, great care should be exercised not to let it bring along special ecological or sociological relationships which may be tied to it by historical chance rather than necessity. For example, it is not obvious on the face of it that sprawling cities are an unavoidable concomitant or that one type of economy is necessary to enable industrialization to arise and function. The estimation of what precisely comes into the valuational base is a complex issue in which economics, sociology and the other social sciences must play a part. Similarly in dealing with central necessary conditions in the domain of political forms it may very well be the case that only some actual democracy can in the long run provide the widespread initiative required to keep going the large-scale collective efforts and enterprises that are themselves an integral part of modern life. This would, if true, give democratic political forms a place in the valuational base, even apart from any further moral quality they may have—for example, their embodiment of respect for individuals as expressing psychological needs.

How central necessary conditions, themselves initially analyzable as means, may function evaluatively to criticize even moral rules and patterns will be illustrated in detail later on.

CRITICAL CONTINGENT FACTORS. Every period has some contingent elements that loom large in applying the valuational base to that period. This is usually because they constitute serious problems or threats to survival or the means of satisfying needs of great value. A threat of war which it seems impossible to avoid may begin to orient all phases of life about it and act as a principle of selection for value decision. For example, when the Athenians saw clearly that the Persians would eventually attack again, after the Persian defeat at Marathon, the menace of Persia became an integral part of the valuational base of that period. The question "Will it help our survival?" could be directed at any social policy or political form or moral rule which might impinge on the issue. Similarly, in a class society in which, for example, a few nobles own most of the land and the vast majority of the people are impoverished, the question "Will it lead to land reform?" may become a central focus to which all evaluation is referred. Or again, physical disasters, such as a great plague, or dessication of the soil, depending on their magnitude, may have a similar effect. If the contingent element is on a large enough scale its demand for consideration permeates to the core of moral deliberation. For example, in our day, many see the problem of population growth outpacing food supply to be such an element. The way in which it poses a problem for those who regard birth control as morally wrong, and imparts an almost moral zeal to advocacy of planned parenthood, shows the ethical implications of contingent elements clearly. While no one ethical answer is predetermined, the fact that a morality cannot ignore a problem is decisive for its place in the valuational base.

The examples given are all contingent elements with reference to the value-system; they need not be values nor means to value which one would choose or welcome. They play their role of determining relevance in ethical judgment by their sheer existence and unavoidable influence, just as in personal life the variable individual facts of disability or old age or illness stubbornly shut off many special areas of choice. Such local elements on the historical scene are not, of course, always historically contingent as they are ethically contingent. They may be among the major determinants of the historical process. It is also to be noted that they play their ethical role because there are values in whose way they stand and which are to be secured only by winning the struggle against the special obstacles. The contingent elements find their place in the valuational base only because these goals—national existence, or the demand for land and greater equality, or health and safety, or the maintenance of a given standard of living—are already located in the valuational base. But it often happens that these contingent elements by their special form will shape the goal, give it its pressing quality, and act as bases of reference in the evaluation process.

§ Validation of the Valuational Base

MAJOR ISSUES.   Up to this point we have been elaborating the concept of the valuational base and giving some indications of its character and constituents. Before we go on to consider more carefully how it would operate in evaluation, a number of issues must be dealt with that concern what we may call roughly the validation of the concept or the logic of its use. It may be asked in the first place, how we would establish that the valuational base does in fact exist. This is a question cast in the realistic language. The parallel in an instrumentalist language would be whether we can specify with some precision the conditions under which the concept may be fruitfully employed. It is not our purpose in this context to decide between such formulations and we may use both more or less indifferently. A second but kindred question is how we should decide whether a given candidate was to fall in the valuational base or outside, that is, the logic of verification involved. A general question running throughout is, of course, the possible criticism of the arbitrary relativist that the concept of the valuational base is only a roundabout way in which ultimate commitments are presented. This needs no special treatment since it is answered by any success in presenting the process of verification and by the consequent assurance that the establishment of the valuational base is not a question-begging procedure.

The answer to these questions is in general a recapitulation of the argument of the several chapters in which scientific contributions were shown to be effective in diminishing indeterminacy in ethical judgment. The bases of indeterminacy were found to lie in extreme field complexity and field instability, making judgments of need-structure and problem-structure almost meaningless. The ultimate validation of the valuational base concept is therefore to be found in the lessons of the human sciences that human life on the various levels of investigation does exhibit some degree of systematic structure.

N.B.

DEGREE OF INTEGRATION AN EMPIRICAL QUESTION. How far the valuational base will constitute an integrated configuration, how far a set of loosely related strands is itself an empirical question. We have already seen that biological and psychological needs take cultural and historical form, and necessary conditions take particular shape in the milieu of contingent factors. Some part of the integration will thus be contributed by the particular character of the period to which the concept is applied.

Even such a question as whether there have been or will be periods in which the concept is utterly inapplicable is one of historical probabilities or scientific prediction concerning the conditions of field instability and field complexity. With respect to the past, the farther back we go in pre-historical times, the more likely are human problems to have been tightly structured by the necessities of immediate survival. In historical times, the best test cases are revolutionary periods, with their surface appearance of arbitrariness and indeterminacy. But it is precisely these—for example, the English revolution of the 17th century or the French revolution—which, on increasing study of their day-to-day progression, show a complex articulated problem structure. Historians, whatever their philosophy of history, tend to fall into the language of historical tasks performed, unfinished tasks, and point to the striking fact that restorations never dare to turn the clock wholly back. With respect to the future, the area of conjecture is greater. It seems likely, however, that given the continuance and development of industrial society the element of conscious planning will sharpen the structure of problem-situations rather than render it more indeterminate. If with a growing abundance the area of individual liberty is expanded, any indeterminacy in individual judgment coming from wealth of choice will not be an ethical indeterminacy, since the ethical principle of such liberty is that the individual has the right to choose whatever he wishes within the given area. If, on the other hand, problems of natural resources and their limitation become central, then the field will be structured by basic necessity. That a multiplication of problems might ensue, more than matching obsolescence of problems and making the whole field unstable through complexity, does remain a logical possibility. But even here, rather than finding the concept of a valuational base inapplicable, we might see it unified by the central theme of a necessity to simplify life.

With respect to the present, the degree of unity to be found in the valuational base for the contemporary world can be read from the general drift of the lessons we have gathered throughout the book. If we think back over the kinds of criteria for evaluation that have emerged, one striking fact stands out: the negative ones have a sharper outline at the present stage of knowledge than the positive ones. By comparison with the fragmentary nature of our knowledge of the good we do have a comprehensive view of evil—whether it be described biologically as death and illness, psychologically as neurosis and the inability to function, socially as wars, depression, mass hysteria, or historically as the decline of a civilization or the thwarting of a people's aspirations. This phenomenon characterizes the scientific concepts that helped fashion the criteria as well. The definition of physical health is just getting to the point where it can be cast positively; that of mental health is still cast in terms of removing conflicts and anxieties. Neurosis is well-mapped, but sublimation is a relatively unexplored area. The cultural criteria of disharmony are more manifest than those of positive well-being and creativeness. The historical thread of growing control over nature stands out more clearly than the human ends which such control can increasingly support. A valuational base for a contemporary ethic is therefore weighted heavily with the effort to overcome known and scientifically describable evils. This effort implies some outline of the necessary conditions of the good, and should lead increasingly to the clarification of the positive good, though like Moses in the desert we may see this promised land only from afar.

How an Item Is Established in the Valuational Base. There is clearly no single answer to the question how an item is established in the valuational base. It depends on the type of item it is. The logic of investigation by which it is established that certain needs are universal suffices to place them in the valuational base. Similarly we saw that some values won their place as perennial aspirations. The mode of verification here was placed in the methodology of analytic historical investigation.

More serious methodological issues arose in dealing with central necessary conditions and critical contingent factors. As the raw material of central problems of an age or people, these presuppose values for whose achievement they are necessary conditions and values that are critically affected by the contingent factors. Now some of these values carry us back to universal needs and perennial desires. But not all do, and even those that do may refer to cultural specializations and particular historical forms. Yet it is not always necessary to track down the values involved. The mark of the degree of integration in the period is often that anyone can write down his own set of values or goals (actual, not simply wishful dreaming, although this might not even have to be excluded) and no matter what he in fact puts down, the central necessary conditions will be required for their achievement and the critical contingent factors will have to be faced. Perhaps not universality, but overwhelming relevance is the test.

Take, as an illustration, the question of a stable peace in the modern world. Many people, differing on all sorts of issues, have come to see it as the primary problem of mankind today. This means that it permeates all areas of life everywhere and calls for decision in terms that will forward its attainment. It is worth repeating that it does not commit everyone to the same plan or the same solution. As a common problem, it does not, for example, eliminate differences of interest. It means that each interest has to formulate its claims for ethical decision in the light of its proposed solution to the problem, and that an interest which is hostile to peace in its effect or which means that men turn away from facing the issue is to that extent less moral, and assumes at the very least, a burden of proof for its own justification. The need for peace thus furnishes one phase for ethical evaluation which may be very powerful without becoming the sole criterion.

Now what is the method of verification by which the problem of attaining a stable peace would be given a place in the valuational base? It would not be a question of savoring the values of peace as against those of war and voting according to taste. Nor would it be simply a question of the general value of survival or life, but a full-length estimate of the impact, character and consequences of modern war throughout the range of human life. It would have to reckon with the psychological views sometimes advanced that war even in modern form corresponds to basic human drives, or serves unavoidable psychological functions; on the whole these are outlooks largely discarded or else referring to the types of war found in the past. It would have to reckon with Malthusian-type arguments on biological overcrowding grounds and with arguments about the selective eugenic benefits of inter-group conflict. It would have to reckon with cultural forms that glorified war. It would not suffice here for the relativist to point out that war may be itself a value arbitrarily chosen instead of peace, giving the Homeric Greek chieftains as evidence. For the issue is modern war and the valuational base in the contemporary world. Fascist doctrine with its glorification of war would be more relevant evidence. But this could not be taken in abstraction. The cases offered would have to be considered in the context of their occurrence. And here the historian might very well show that the glorification of war in Fascist and Nazi usage was chiefly instrumental, partly to solve internal political and economic problems and partly to foster a deliberate irrationalism in the attack upon both liberal institutions and socialist threats.

Finally, the procedure of verification would have to show that war in its modern form does have a devastating impact on all concerned, whatever their actual different values, that it is fatal to victor and vanquished alike, that it provides no solution for the problems that drive nations into war, and so forth.

An even stronger dimension of verification could be worked out by analogy to the theory of needs in psychology. Whether it would apply in social and historical issues would be an empirical issue; it might apply in some and not in others. Let us see what it would mean in the question of peace. A need is characterized by its unsuppressible disturbing power so that where not expressed satisfactorily it drives the individual into some distorting pattern of expression, whether stable or unstable. On this analogy, it would have to be shown that the problem of peace or war in fact becomes an issue whether men realize it or not in all that they do. Hence it would be argued that they are in some sense today constantly facing the issue directly or indirectly by the attitudes they adopt, the actions they perform or leave undone, and so on. Hence it involves the hypothesis that any advocacy of an alterative to peace in this historical period represents not a genuine value but either an *indirect* way of facing the *same* problem on different factual assumptions (for example, that a quick short war would be the best way of bringing stable peace), or a running away from the problem, or camouflaging even to oneself special interests that cannot stand open scrutiny. On such a hypothesis, not only glorification of war but even a pessimism which says man is going to perdition and a wise person will therefore ignore the question, or an insistence on acting on abstract principle irrespective of the cost to mankind, will be held to be, *in the actual contexts in which they occur*, retreats from the problem itself, rigidities serving some limited functional role of interest which cannot itself bear the weight of formulation as a moral theory.

Perhaps even so important an issue as peace has not achieved the status indicated by this needs-model, in which the specific demand permeates all life and is relevant to every item of behavior. Nevertheless it is possible that life at some times in some historical situations has such a tight integration, and it is important to see that we are dealing with a logic of verification that is capable of empirical formulation.

THE TRIALS OF A CANDIDATE—THE "BROTHERHOOD OF MAN." It might be said that peace was able to establish its place in the valuational base because it had so many testable *instrumental* relationships. What would a traditional and venerable *ideal* do to win admission to this select club? Suppose the ideal of the brotherhood of man applied and sent in its credentials. It might state that it had excellent theological and philosophical references, and from differing points of view. The Christian outlook calls it a consequence of the fatherhood of God. The Stoics thought of it as the divine fire in all men and preached a universalist cosmopolitanism. The French revolutionary philosophers exalted it as fraternity, and the Marxian materialists look forward to it as the classless society. An ancient philosopher could still say, as Aristotle does in his *Eudemian Ethics*, that the controversy whether the best life is that of the intellect, of ambition, or of pleasure, does not concern slaves and those who spend their lives in menial pursuits. But Kant addressed his ethical commandments to every rational being, even any rational non-men there might be anywhere in the universe. And no serious moral philosopher has since then tried to work out moral questions for anything less than human beings generally. Individualism, no matter how extreme, casts its morality as the expression of the individual nature of *any* man. Often, as in Utilitarian political theory, it has shed its natural rights claims and asked for its liberties because to maintain them will in the long run yield the greatest happiness of the greatest number. In some ethical approaches today the frontier of criticism has even shifted to the extent of finding the human reference too narrow. Though ethical attitudes extended to the animal world have often been justified in terms of effects on human character, Albert Schweitzer's conception of reverence for life orients one's moral outlook to the living as such, not simply to the human.

Are these references sufficient to admit the brotherhood of man to the valuational base? The opposition too will have its say. "Ideals," it will charge, "are many and relatively easy-going. To take them in will lower the quality of the club. It will no longer be an effective organization for evaluating concretely social policies and working moral attitudes, but a club for idle wishes and hopeful dreams. Let in one ideal of this sort, and you can't keep out its friends. Before you know it, there will be liberty and equality and who knows what else. In any case, the credentials presented are one-sided. What kind of a recommendation would the brotherhood of man get from the philosophers of egoism and nationalism, or the various shades of élitism? For that matter, even the practice of its proponents often shows scant respect for the ideal they recommend. Christian peoples have sometimes found the brotherhood of man in relation to God quite compatible with the slavery of man before man. In retrospect, aren't the French revolutionary ideals just the universalistic pretensions of the middle class demanding its place and power? And the Marxians push their classless society into the future and meanwhile insist on judging from the point of view of the proletariat or the interests of the Communist party. No, no sensible realistic person can say that the brotherhood of man belongs in the base of effective valuation today."

Before accepting or denying the application for membership, the ideal of the brotherhood of man would itself have to be analyzed and clarified. Obviously its reference to the brother relation implies a familial context, and the appropriate structure of the family it envisages would have to be articulated. Without undertaking this task, let us here deal

with it in a very minimal sense. Let it consist in the specification that the group to be reckoned in a contemporary
valuational base is all mankind, not any restricted sub-group
such as nation, race, class, or lone self. In short, it becomes
the minimal assertion that every one is *to count* or *to be
reckoned* in the estimation of the good which determines
policy. The application of the ideal of the brotherhood of
man to the valuational base may thus be reformulated as
a hypothesis that the valuational base in the contemporary
world will have a *global scope*.

Even this clarification and limitation does not carry us
far enough. There may be all sorts of loose ends in the fundamental ideas. For example, the time reference needs some
amplification. Is the group reference to the living or does it
include future generations to some degree? Thus moral attitudes to conservation of natural resources—the contemporary
analogue of the virtue of thrift and the vice of prodigality—
will involve asking directly whether we are ready to rob
the future generations for the sake of the present. And
deliberation on it will include estimates of probable scientific
discovery of substitutes, not simply "What do we care about
them?" But letting go any further refinement of the concept
of global scope and thinking of it chiefly in contrast to more
restricted forms, we may see in line with the major thesis
of the book that such questions cannot be settled by stipulation, by simple introspection on the state of our feelings
about mankind, nor on the other hand by a direct set of
observations which take no account of value-attitudes. The
conclusion that the valuational base in the contemporary
world is characterized by a global scope would have to rest
on an analysis of the interplay of ends and means, needs
and instruments, ideals and practical possibilities, in the
whole structure of the modern world. The case for the
hypothesis would involve at least the following tasks, embodying theoretical analysis, scientific data and specific
valuations.

(1) A critique of egoism would have to be carried out,
which might prove as extensive as our whole treatment of

ethical relativity. A number of threads in this critique have already been developed at various points of the book. It was seen that the view that a man should pursue his own interest has often been given a social formulation—it is believed it will lead to greater initiative and productivity. Again, a great part of what functions practically as individualism was seen to consist of *permissible* differences already justified implicitly on a social ethical standard—for example, in the cultural relativist stress on tolerance. This need not be an arbitrary tolerance, since it, like the theory of individual liberty, is capable of elaboration as a systematic account of grounds, limits, and occasions where cultural selection or personal interest constitutes a justifiable decisive factor.

In addition, our inquiries yielded a considerable devaluation of the psychological bases for the individualist formulation. We saw, in examining fundamental biopsychic valuations that probably no man's death or pain would be a real end-value to another, although this did not preclude one man's using another's life or pain to further his own life and pleasure. Similarly, in the psychological perspective we saw that there were universal needs, though some might satisfy their needs at the expense of others. The psychological treatment of egoism and our consideration of cultural and historical perspectives suggested that self-asserting self-aggrandizing patterns might be found evil in terms of their psychological, social and historical consequences. Thus implicitly to take the standpoint of the many against the one requires further justification.

This is found in the further fact that both psychologically, culturally and historically, morality is an interpersonal and social phenomenon. It is social in precisely the same sense that language is social. The growth of the self is a process of "socialization." Moral patterns are socially caused, and function socially. Egoistic ethical theories, when historically studied, are themselves found to express group aims and special interests rather than the lone appeal of a man that he has but one life to live. Even this last argument when seriously advanced usually ignores the problem of the kind of self that is his, treating it instead as a given atomistic unit with an inborn "selfish" character.

Such consideration converges upon the belief that the valuational base in the contemporary world requires a social rather than egoistic-individual formulation of moral issues. The decision is one of methodological policy, justified by its consonance with the trends of knowledge indicated, and by a further estimate of the consequences of the alternative formulations in the modern world. (The local reference is to be noted; the same conclusion might be reachable with regard to other times, but it would require in part independent examination of evidence.) These comments have constituted not a demonstration, but a bare suggestion of an evaluative task.

(2) A critique would also be required of any view that insisted on a more restricted group reference than all mankind. There have been many such views—for example, that moral judgment ultimately is to be made from the point of view of a family, a class, a nation, a race, an élite. The general lines of such a critique would be threefold. It could be shown that many of the specific interests of these lesser groupings can readily find a place in a global moral structure without defining its scope in their own limited terms. Thus one could probably readily justify a man's closer moral bonds to his family than to strangers; or each man's having the moral obligation to support his own wife and children today rather than his sister and her children, leaving the support of his wife and her children to her brother, a system found in the Trobriand islands. Similarly, the moral proposition that a group having cultural and linguistic interests in common and desiring a national form of life ought not to be interfered with in its formation and exercise is probably now more firmly established on a global ethics than on a narrower national base. In the second place, a critique could be carried out of many of the factual claims offered to support the various more restricted perspectives. Thus the pseudo-biological pretensions of race superiority, the theses

of male superiority, the assertions of peculiar national des-
tinies to rule, or élitist claims to superior political wisdom,
have failed to run the gauntlet of scientific scrutiny. In the
third place, a careful examination of the theoretical arguments in some of the more restricted conceptions will show
that they regard their restrictions either as instrumental to
wider development or else as necessary because anything
wider is practically unachievable or utopian. For example,
the Marxian thesis that in the modern world ethical judgment is to be made from the point of view of the proletarian
class rests on the claim that the interests of the proletariat
are the emerging interests of all mankind. Again, the defense
of the national interest as the basis of judgment today is cast
by Kennan as a claim that our national interest is all we can
really know and understand, but is coupled with the hope
that decent purposes and undertakings in the national interest can never fail to be conducive to a better world.[5] The
conclusion of this third line of analysis might very well be
that in the contemporary world evaluation of policy and
ethical judgment may better be cast directly in terms of
global welfare.

The more positive analysis of the restricted approaches
would consist in showing that in the contemporary world
no definition of lesser group interest and specification of
mode of attainment is possible without assumptions that
tend more and more to involve wider relationships approxi-
mating global ones. The hypothesis of a global scope for a
contemporary valuational base thus accepts the view that
there are major interests of every group that can now be
realized only on a global basis. Even though some groups
may see these purely as instrumental relations—say a nation
reckoning other nations only in terms of its own advantages
—the effective relationships are such as to force increasingly
a whole global mode of reckoning. (This is reflected in contemporary controversy where the aim to safeguard national

---

5. George F. Kennan, *American Diplomacy 1900-1950* (Chicago, University of Chicago Press, 1951), p. 103.

interests *no matter what the cost to the rest of the world* has rapidly lost its nobility.) And the consequences of such a mode of reckoning are important for ethical judgment—for example, in the clear branding of genocide as a crime, in the moral rejection of colonialism, in the recognition of national independence of other nations as a moral claim, and in hosts of specific condemnations of discriminatory practices.

(3) A third type of inquiry in establishing the brotherhood of man in the valuational base would lie in showing how the convergence of contemporary problems and forces have given this ideal a central significant role. Part of this might relate the ideal to modern modes of life and the effect of central necessary conditions—how the growth of industrial society increasingly makes available a common mode of life all over the globe which provides a basic participation and common roles for the vast majority of the people and brings them together in specific ways that genrate a common outlook. Part of the inquiry would relate to the contingent historical picture—the defeat of Nazi racialism, by far the most devastating in the history of the globe, the breakdown of century old colonialisms in Asia and Africa, the growing pressure for the removal of discrimination—whether racial, religious, class, sex, throughout the whole human scene. The convergence of these multiple forces would explain both why the ideal of human brotherhood and dignity for all people has gathered such terrific strength in the modern world and why it forces itself on consciousness as a central moral task.

(4) The fourth task would be directed not to critiques of opposing conceptions and exhibition of grounds for the growing strength of the ideal of the brotherhood of man, but to articulating the ideal itself and tracing its value relations. It would exhibit its connection to universal needs and to desirable psychological attitudes in interpersonal relations. It would explore the intrinsic values in both the general brotherhood relation of men and the specific forms such a relation might take in the modern world. It would show how acceptance of the ideal would serve to resolve the problems

which give it its central contemporary relevance. Such a general evaluation of both intrinsic and instrumental elements would trace implications in major areas of personal and social life, in qualities of consciousness as well as social relations, in institutions of cooperation as well as structure of sentiments.

There are no doubt numerous issues of fact, of analysis and of specific valuation in the program we have outlined for deciding whether a global reference today belongs in the valuational base. But the answers need not be arbitrary: the issues of fact are determinable and many already have been settled, the analyses can be accomplished with patience and persistence, and many of the specific valuations prove to be quite determinate. And if biasses turn up in the analysis, they too can be assessed for their assumptions and their role, and thus minimized.

That the brotherhood of man in the sense of global scope will turn out to have a high positive value seems to me fairly clear. Whether it will turn out to have the deep roots and multiple ramifications that a place in the valuational base demands is a more difficult issue. My own belief is that it does, because it is tied both to fundamental needs, perennial aspirations and major goals, central necessary conditions, and extremely critical contingent factors. But this judgment constitutes a prediction as to the general outcome of the types of inquiries outlined.

§ The Valuational Base
in the Evaluative Process:
Illustration of a Permanent Instrumentality

MULTIPLE MODES OF EVALUATIVE FUNCTIONING. We have seen that the valuational base has a wide scope in evaluative processes. Its impact is felt on decisions of social policy, in the critique and reconstruction of moral attitudes and ethical outlook. Its effect is secured not in one way but in modes of functioning corresponding to the types of constituents in the base. The examples of peace and global scope discussed above show how considerations of universal needs, focal objectives, central instrumentalities and critical contingent factors become intertwined in the validation which determines the inclusion of an item in the base. And once an item is included its modes of functioning in further evaluation are likewise multiple. Thus the need for peace acts as a basis for evaluating types of character to be fashioned in the modern world, as well as desirable social policy, and the demand for a global scope becomes the basis for a reconstruction which cuts through all lesser and restrictive moralities. These are continuing ethical tasks in the modern world, far from finished. In numerous ways as yet unreduced to systematic classification, the valuational base provides standards of varying degree of specificity.

Perhaps the best way to see how it functions evaluatively is to follow out examples in greater detail. Let us take one of a major permanent instrumentality and one of a complex ideal in which the fusion of means and ends is especially apparent. Raising the productive level of the globe is perhaps the best instance of the first, and expanding democracy the most ambitious instance of the second.

WHY RAISING THE PRODUCTIVE LEVEL OF THE GLOBE BELONGS IN THE VALUATIONAL BASE. Productive power was seen to be a major aspiration in our historical core-constellation of values. In this general sense, we expect always to find its maintenance in the valuational base, and its mode of functioning is of special interest as an example of a permanent instrumentality or central necessary condition. In the contemporary world, however, it takes a special form. It is no longer an isolated individual aim, nor a purely national aim. It appears already in consciousness as the effort to raise the productive level of the whole globe, including the industrialization of many undeveloped areas and the systematic ushering in of the atomic age. It is recognized as a necessary condition for many human goods. For it will bring tremendously increased opportunities for the satisfaction of human material needs, as well as opportunities for expansion of knowledge and cultural development. In addition it has wide ramifications through many of the problems of conflict which constitute critical issues in the modern world. There is accordingly little difficulty in seeing the task of raising global productivity as an item in the valuational base.

This conclusion is strengthened when we turn to the moral relations of the task of expanding the productive level, especially when it takes the form of substituting the prospect of abundance for the existence of scarcity. I have discussed this problem elsewhere in greater detail both in its causal and evaluative aspects,[6] and will therefore simply recapitulate the thesis briefly and dwell on some of its implications.

The underlying thesis is that the fact of scarcity has played a greater role in ethical theory and moral patterns than is commonly recognized. In an era of low productive capacity scarcity has appeared in human consciousness chiefly as the unreasonable demands of the passions and the appetites. Therefore morality has on the whole had a restrictive outlook. This characterized not only the moral man seeking constantly to be "master of himself," but also interpersonal and social relations. Property concepts remained essentially negative—holding one's own against encroachments. Concepts of social control were cast in terms of anticipated conflict and dispute, and the ideal of harmony as minimizing conflict.

Increasing abundance makes it possible to remove many bases of conflict. Developed in a complex social process, it admits of wider planning, greater forethought and increased participation. It requires no submerged class who shall not share equally in the benefits of industry. The result is a reorientation of outlook towards social welfare, which affects obligations and principles of social justice, concepts

6. See my "Scarcity and Abundance in Ethical Theory," in *Freedom and Reason,* edited by Baron, Nagel and Pinson (Glencoe, Illinois, The Free Press, 1951).

of social control, as well as fundamental notions of property, liberty and individual rights.

REORIENTATION IN THEORY OF OBLIGATIONS AND VIRTUES. Whatever other modes of evaluation are employed for obligations and virtues, an additional dimension stems from their social role in relation to developing and maintaining abundance. Just as the Puritan virtues of diligence, abstinence, sobriety, seriousness, thrift, played a part in the earlier history of economic advance, so in a more self-conscious way there is now a gearing of obligations to production. Peoples seeking to develop industrialization recognize the necessity for some changes in habits and outlook.[7] Where a complex technology already exists, the maintenance of production and its services assumes an almost moral character. All sorts of economic policies and taxation policies are felt to be justified in terms of maintaining and developing industry. All sorts of regulation of standards of quality are imposed as requirements of public welfare. Traditional private claims give way when the issue is one of developing major new resources. And not merely public duties but even phases of personal character are shaped into virtues and vices by the growing needs. For example, specific traits such as punctuality and attentiveness, or decisiveness, begin to take on a moral quality; and in general, the sense of responsibility becomes a moral requirement where failure in a limited area may cause great damage or bring the whole process to a stop. While abuses may arise in this area in the confusion of economic policy and fundamental morality, yet the over-all character of the effort to keep production going and developing is seen as increasingly moral precisely because it concerns a core means to whole constellations of human goods. Thus it serves to moor obligations all the way from the realm of personal habits to public decisions of policy.

REORIENTATION IN PRINCIPLES OF SOCIAL JUSTICE. Principles of social justice concern distribution of rewards and

---

7. See above, p. 223.

burdens. Moral principles for rewards have generally been
cast in terms of some combination of need or work or incentive, the combination varying with type of society. The
meaning of need has varied in different social theories. The
sense of impecunious is one extreme; the needy are taken
care of by private or religious charity. The use of need as
the basic principle of assignment is another—for example,
in the communist ideal of "to each according to his needs."
Amount and kind of work done has likewise been used as a
principle of distribution in part at least in most systems,
though it may be subordinated in some kind of status system
where distribution is based on a position in a hierarchy. In
a laissez-faire economy which formulates its aim as maximum encouragement of individual initiative and thereby
productivity, the encouragement of incentive both to work
effort and to investment has a large role among the principles guiding division of the returns of production.

As the realistic possibility of abundance grows greater,
it imposes an emerging moral pattern on principles of reward. Work remains a central basis—we need not even in
the present context raise the question why a man should get
a return for laboring. (The chief bone of contention between capitalist and socialist theory on this point has been
whether the laborer does get a "full" return or only a "partial" return in capitalist society.) Incentive has tended to
lose much of its moral quality with the limitation of laissez-
faire. It is asked why people need *extra* rather than normal
reward to exercise abilities which are socially desirable.[8]
Perhaps judgments about incentives should be made piece-
meal, particularized for different fields. There is obviously
considerable difference between inducing men to engage in
hard unpleasant work, and inducing them to become movie

---

8. Similar questions are raised about return on investments in the
theory of ownership. Thus A. A. Berle and G. C. Means, in *The Modern
Corporation and Private Property* (New York, etc., Commerce Clearing
House, Inc., 1932), questioned whether risk without management really
deserved the kind of returns that were traditional for owners who exercised
control over their wealth (Book IV, ch. 1).

stars or artists whose work is itself gratifying. Again, a great deal of the answer to the moral status of incentives obviously depends on the picture that social psychology may give us of human motivations in different social milieus.

What most strikingly characterizes the restructuring of principles of distribution is the emergence of need to a central place. It becomes stabilized at the very least as a moral conception of a floor below which no one will be allowed to sink, no matter what his incapacities. It begins to penetrate the reward-for-work principle through all sorts of social devices in the lightening of burdens—family allowances in one place, income tax deductions in another, rent for living-quarters expressed as a percentage of income in a third, social security in all. When there is abundance in a particular field, wider needs are met through cheaper prices in a competitive economy, through public distribution supported by taxation in a socialist economy. Emphasis on meeting needs is seen in the role of the notion of consumption in economic analysis. Consumption becomes part of the process by which the wheels are kept going, and its relation to human needs and their satisfaction becomes pivotal in the whole process. Thus if instead of consumption we have destruction of products to keep up prices, or if we have diversion to products that admit of "consumption" only in a spurious sense, such as diversion in the now classical phrase from butter to guns, there is obviously something *wrong*. In our own day this "wrong" has already acquired a moral sense, and this reflects the demands and expectations that have accompanied the growth of the technological promise of abundance.

Strange as it may seem, therefore, one of the most significant points of large-scale moral decision is the point at which the ratio of consumer goods to producer goods is determined. In undeveloped countries, the decision to make vast sacrifices for industrial development is complicated by the fact that sacrifices are made by one generation and the fruits will be gathered by another. A global plan of global development consciously geared to minimize the hardships of all peoples involved would therefore have a high moral quality. On a smaller scale, United Nations agencies such as the World Health Organization, the Food and Agricultural Organization, the United Nations Educational Scientific and Cultural Organization in its applied work, embody the same moral standpoint. These are the terms in which social justice becomes cast when abundance is in sight, and becomes a sufficiently realistic goal to act as one of the moorings for morals.

REORIENTATION OF CONCEPTS OF SOCIAL CONTROL. Concepts of social control assume a more functional character with growing abundance. A large and complex social organization thinks more in terms of guiding individuals to beneficial social roles than coercion of trespassers. Law becomes geared to the various major human enterprises and metaphors of engineering spring readily to mind. Property concepts shift to the positive social idea of organizing resources for production, as the categories of small-scale ownership wear thin in dealing with basic natural resources and the wielding of atomic energy.

Growing possibilities for abundance also affect the further ideals which determine what to do with the increased productivity. Thus liberty itself, for example, becomes the positive ideal of providing concrete opportunities for human development. In general, individual rights tend to be cast more as social opportunities for individuals, and individual duties to be supported by a structure of general social assumption of major responsibilities.

Such changes have, of course, by no means taken place in most moral patterns as yet. But they are further advanced than one might have expected in ethical theory and in the analysis of ideals and responsibility in social philosophy. And they are relevant today, even where scarcity still exists; for the shift in attitude entailed by the prospect of abundance serves itself to provide a standard which helps it along. Moreover, even the possibility so stressed by some that abundance may be limited by pressures of population and depletion of natural resources need not necessarily have a reversing moral effect. On the contrary such possibilities by their focus on the struggle for abundance as against the encroachment of scarcity might intensify the critical role of this component in the valuational base.

§ The Valuational Base
in the Evaluative Process:
Illustration of a Complex Ideal

THE PLACE OF EXTENDING DEMOCRACY AS AN ITEM IN THE VALUATIONAL BASE. Many of the necessary conditions as well as perennial values in the modern world have become integrated in the ideal of democracy as a way of life. The view that a central task of the modern world is the extension of democracy is a common one, and the inclusion of this task in the valuational base applied to the contemporary world appears to be warranted as an hypothesis. Its validation would go through many realms of personal relations and social structure. We cannot begin to carry out this analysis in the present context. But on the whole the validation is likely to be conceded—for we are accustomed to regarding democracy as a moral outlook—and the really crucial issues thus tend to be shifted within the concept of democracy itself. On the one hand there is the tendency to narrow democracy to a purely political form, on the other hand the tendency to subordinate political form to the social benefits and their spread to large masses of men. Thus a recent UNESCO symposium on democracy notes repeatedly the tendency to formulate western vs. Soviet conceptions of democracy as political vs. social.[9] There has also been an unfortunate tendency to analyze the issue as itself emotive or persuasive, stemming from the conflict of the great powers. This overlooks the profound theoretical questions that have to be faced in fashioning a concept of democracy

---

9. *Democracy in a World of Tensions,* edited by Richard McKeon (Chicago, University of Chicago Press, 1951).

adequate to serve as a constituent of the valuational base in the contemporary world.

It should not be assumed a priori that the conflicting conceptions are each talking about something else, or that they represent ultimate cleavages in value. They may be stressing different lessons in the experience of mankind—for example, the one that no group however well-intentioned can be trusted with absolute power without effective political controls, the other that no political mechanisms however perfect in form can fulfil men's aspirations when economic power is concentrated in the hands of a minority. Again, there may even be a common framework underlying the differences. For example, McKeon, criticizing the tendency to oppose the two, maintains that a common conception of "rule of the people in their own interest" underlies the varying uses.[10] It is interesting to note that classical liberalism in the 19th century found no difficulty in looking beyond the political forms to the conditions under which they might be effective and the goals they might achieve. J. S. Mill stressed the type of character that the principles of political liberty were intended to produce. The educative role of political institutions is a fundamental point in his *Representative Government* and the importance of an independent self-determining character to human progress is the underlying theme of his essay *On Liberty*. Even those who raised the cry of liberty versus equality quite clearly thought in terms of the type of life and society that would prevail under the respective formulae.

It becomes necessary therefore to go beyond initial opposing definitions of democracy to the theory of history or society or psychology on which the alternative positions are based. If political democracy—in its origins, development and intended functioning—has always meant to the mass of men who supported it the hope of extending a higher standard of living and the wider development of human capacities, then there is no reason why such a promise should not

---

10. *Ibid.*, pp. 195-196.

be regarded as part of the very goal content of democracy. Or if opposing conceptions of democracy rest on different views of the way in which social change comes about, or different assumptions of the nature and capacities for government of the mass of mankind, such beliefs too have to be investigated in clarifying and stabilizing a contemporary conception of democracy.

ELEMENTS IN THE CONCEPTION OF DEMOCRACY. The task of extending democracy in the contemporary world does not presuppose a complete answer to these issues. It can begin the work of evaluating using negative criteria of extreme lack of democracy upon which it can find agreement; for example, it can point to surviving feudal forms and the recent Nazi and Fascist forms. It can utilize partial criteria of literacy, vote and participation in government, and removal of external domination, as the Trusteeship Council of the United Nations has used in removing countries from colonial and dependency status. Impartial studies may begin to be possible in practice in some areas, as they are already possible in principle, of the extent, degree and quality of political participation by the mass of the people, using the criteria of the educative and initiative-provoking role of representative institutions. And while one does not find formal agreement on such a conception as the right to work, there is growing recognition of material welfare minima and governmental responsibilities in these respects, whether achieved through competitive or collective enterprise and whether they are seen as part of the definition of a democracy or as a proper job for a democratic government. On the whole, while one may not minimize the basic disagreements that may exist in fact between nations and classes in the modern world, there is no ground for despair in the task of elaborating a stable moral core-conception of democracy.

In the light of its historical development, a full conception of democracy would involve, to summarize briefly, major goals or ideals, such as liberty, equality, fraternity, the dignity and development of the individual. It would involve political principles for giving operative meaning to the concept of the will of the people. And it would involve —as every profound theorist or student of democracy from Pericles to Jefferson and de Tocqueville and up to present social philosophy has made clear—habits and attitudes and modes of thought consonant with its goals. It would involve an economic life in which social objectives are controlling and all who are affected participate in the shaping of policy. It would involve the ideal of an educational system which enlists the full initiative and resources of all its participants, and of a family life which embodies shared values and responsibilities rather than peremptory paternal command and familial obedience. And similarly for all other institutions.

Carrying out a democratic moral critique of our institutions will deepen our insight into the values of interpersonal and social relationships that a democratic life can bring. The actual achievement of democracy is not the task of morality alone. But the need for its achievement remains a stable base for moral evaluation.

DEMOCRATIZING MORALITY ITSELF. The necessity for extending democracy in the modern world enjoins the task of democratizing morality itself. This is a far-flung enterprise once we realize how deeply traditional morality has been authoritarian. It means, as is often pointed out today, incorporating in moral processes and methods the spirit of scientific inquiry. This involves a greater readiness to change and learn from experience, a deeper search for the bases and consequences of moral beliefs and attitudes, a cooperative spirit rather than one of peremptory order and obedience. It may even involve penetrating into the sense of morality itself and softening its guilt-ridden authoritarian feeling. And all this must be done without weakening the decisiveness of moral judgment and the sense of individual responsibility inherent in proper democratic functioning. It is a difficult but by no means an impossible task.

§ The Concept of a Personal Base

A Small-Scale Model of the Valuational Base. A valuational base provides the basis for diminishing indeterminacy in the general ethical judgments which men make about their world and their obligations in it. When they turn, however, as individuals to apply these judgments to particular situations and particular persons—especially to themselves individually—the task of application proves more than simply an act of subsuming a particular under a universal. The same kind of analysis which the concept of the valuational base made possible on the wider field has to be continued in the narrower field. In dealing with the individual we find a concept somewhat analogous to the valuational base playing a parallel role. Let us call it a *personal base*. It includes not only values but also individual characteristics as they have taken shape in his own psychological and social development. It thus provides a frame of reference for value decisions concerning him. Like the valuational base, the personal base will contain universal needs of which the individual may or may not be wholly conscious. It will contain perennial goals. It will have reference to some degree to the central necessary conditions of the social valuational base. It will include large contingent elements that are a function of the specific life experience of the person—from individual physical and intellectual endowment to chance factors that have a molding influence. The personal base, thus understood, is a small-scale model of the valuational base. It is permeated by history too—the individual history of the person—and constitutes his value frame of reference. Not that he always refers to it; for it is not simply his "interests" as he feels them. It is more like the "self" which he looks for when he seeks to know himself.

Validation Procedure. The procedure for validating such a concept is comparable to that for validating the concept of the valuational base. It is not a matter of definition, but of showing that every individual in fact has such. This is a heuristic principle embodying what appear to be the lessons of biology, psychology, literary history, and especially biography as a historical study oriented to value description and analysis. Inquiry into differentiating factors that may have ethical relevance is still in its early stages. Spranger attempted in relatively surface terms to distinguish types of individuality by reference to basic attitudes, such as theoretic, economic, aesthetic, social, political, religious.[11] Charles Morris has looked for correlation of preferred paths of life with physical types.[12] There is increasing study of individual reactions on the biological level,[13] and of personality differences as manifested through projective tests. The growth of psychological knowledge and increased attention to social context have provided more powerful tools for the study of individual development over a whole life span. This is especially evident in the problems of biography.

CONSEQUENCES IN DIMINISHING ETHICAL INDETERMINACY. A realization of the role of the personal base enables us to lessen further the indeterminacy of ethical judgment on the individual level. As Aristotle pointed out, men pray for the good, but they should pray that what is good in general should also be good for them as individuals. Most of the invariants we saw established in the different perspectives were phase rules, elements always to be reckoned in the evaluation of a situation. For the individual, his personal base plays the role of a principle of relevance, helping determine which conclusions reached for a society or for mankind are pertinent to him.

---

11. Edward Spranger, *Types of Men* (5th ed., trans. by Paul J. W. Pigors, Halle, 1928).

12. See his "Individual Differences and Cultural Patterns," in *Personality in Nature, Society, and Culture,* ed. by Clyde Kluckhohn and Henry A. Murray (New York, Knopf, 1948), pp. 131-143.

13. See, for example, the account of preferred behavior in posture, motor processes, etc., with reference to individual as well as species constants, in Kurt Goldstein, *The Organism* (American Book Company, 1939), pp. 340-366.

§ Ethical Theory Itself Moored to the Valuational Base

FUNCTIONAL ROLE OF ETHICAL THEORY. The more deeply evaluation of particulars (in an age, country, individual) becomes rooted in a fusion of universal and local elements relating valuation and existence, the more we may expect protests from ethical theory, which has traditionally had the tendency to transcend existence and judge in regal purity from above. This tendency is common to both supernaturalist ethics and relativist ethics, although the point of reference of transcendence is different. The case for a valuational base is therefore unfinished until ethical theory is also moored to the base.

To some extent this has already been done at various points along the way. We noted at the very outset in presenting the functional approach of biological evolution that the evolutionary perspective will not allow a questioner to step completely outside of the evolutionary process. His standards, aims and strivings, concepts and theoretical frameworks, themselves have to be examined to see what strands, forces, directions, problems in the evolutionary process they represented, expressed or faced. Similarly, we saw how moral patterns were analyzed from a cultural point of view in terms of their functioning as part of the whole cultural process. Ethical theory is only one step removed from morality in this respect; the intellectual tools for moral construction and reconstruction are part and parcel of the same cultural milieu. And the widest view of an ethical theory is found in its historical career, including its historical relation to various fields of knowledge as well as its participation in the way men face historical problems.

The nature of ethical theory is part and parcel of the general question of the nature of consciousness and of intellectual production. This is not, it must be stressed, a question that ethics can decide within itself. The nature of consciousness and the nature of intellectual production are scientific questions in the field of psychology, social science and history. And here the trend of evidence seems against our transcendent-minded purist. Rather, it is probable that an ethical theory will have some historical content and historical functions as well as historical causation.

Fuller confirmation of this thesis will come eventually from the history of ethical theory. On the whole, hitherto, the ethical doctrines of the great philosophers have been studied out of context and more with relation to the analytical tools that can be borrowed from them for use in other contexts. Yet every study of their relation to the major problems of their own time suggests increasingly that not only their own application of their ethics but even their fundamental theoretical concepts have deep historical roots.[14]

THE EVALUATION OF AN ETHICAL THEORY.   The evaluation of an ethical theory requires corresponding expansion. As we have seen, an ethical outlook can be referred to the valuational base, both in its universal and its special elements. Thus an ethical theory can be evaluated for its contribution to the growth of knowledge or its adherence to the best available knowledge in its picture of man and society. It can be evaluated for its conceptual refinement and the degree to which it fashions conceptual tools for dealing powerfully with problems in its valuational base. It can be evaluated by the degree to which it expresses and improves the formulation of valuations—universal and persistent as well as specialized and local. And so it can also be evaluated for its relevance to the central efforts of the era—whether it avoids the problems, whether it throws obstacles in the way of their solution, whether it distracts attention, whether it steers men to facing or avoiding the

14. See, for example, Lewis Feuer's analysis of so apparently remote a figure as Spinoza, in "The Social Motivation of Spinoza's Thought," Proceedings of the XIth International Congress of Philosophy, Vol. XIII, 1953, History of Modern and Contemporary Philosophy, pp. 36-42. For suggestions on the historical dimensions of pleasure theory see my "Coordinates of Criticism in Ethical Theory," *Philosophy and Phenomenological Research*, VII: 543-577 (1947), esp. 554-559.

issues, whether it helps point to successful resolution, and so forth.

ETHICAL THEORY AND A CONTEMPORARY VALUATIONAL BASE. It is an implication in this analysis that some decisions in ethical theory itself are properly tied to the moorings furnished in the valuational base. This should be rendered explicit.

A fundamental instance of this, not always recognized, is the acceptance of knowledge as the criterion in an ethical analysis involving material concepts. For example, an ethics using the idea of pleasure, or desire, or loyalty, or conscience, is committed to accepting whatever authentic knowledge there may be about these phenomena. It cannot simply ignore what we have come to know about such entities or processes, or if it does, it thereby discredits itself. We saw above that knowledge was a persistent element in the valuational base; such acceptance of it as a matter of course is therefore not surprising.

Specific decisions, even of a high order of abstraction, may follow in the light of the more particular content of a contemporary valuational base. It is often recognized in ethical theory that there is a choice between theoretical constructs which has to be made on "pragmatic" grounds. Thus one type of definition of "good" may be preferred over another because it will encourage reflective appraisal in judgments employing the term; or because it will embody a group standpoint rather than an individual standpoint; or because it will provide an intersubjective analysis rather than a subjective one, and therefore make agreement more probable. Now how would we justify each of these "preferences" over its opposite? The process of justification leads eventually to the items in the base—for example, the need and aspiration to expand systematic knowledge, and even the need to reduce conflict and secure wider agreement.

Such valuational grounds for decision in ethical theory would have to be explored in considerable detail in a full-scale comparative analysis. The question, it is important to stress, is not one of substituting valuational criteria for other criteria of some theoretical sort. It involves rather the recognition that some theoretical decisions do have a valuational or pragmatic component, and that this may better be rendered explicit.[15]

SOME METHODOLOGICAL CONCLUSIONS. It is perhaps desirable at this point to underscore some of the methodological conclusions about evaluation that have emerged from our account. These comments are prompted by the apparent circularity in having some components in ethical theory determined by reference to the valuational base, while the valuational base itself was reached in part by an evaluative process using ethical concepts. It was suggested above[16] that ethical inquiry might turn out to have the same type of non-vicious broad circular character that many philosophers of science ascribe to scientific inquiry. And this is the character that on the whole I should be inclined to assign to the process of evaluation in which the valuational base is established.

The valuational base on this view does not consist of value axioms antecedently given, but of principles and policies that are supported by knowledge and valuational experience. These principles and policies are amenable to refinement and alteration in the light of further growth in knowledge and extended valuational experience. Even universal needs do not automatically acquire a value character just because they are universal and are included in the base. As we saw, contingent obstacles of wide scope in the base may actually stem from evils to be avoided, and in principle a universal need might have this character. (We saw, for example, that if there were aggressive needs they would have to be reckoned with, and so would be included in the base but not glorified as a consequence.)

On the other hand, valuational experience need not con-

N B

15. For fuller analysis of valuational criteria in theoretical decision in ethics, see my "Coordinates of Criticism in Ethical Theory," op. cit., especially pp. 559-577.

16. P. 74.

sist of isolated self-certifying value data. There may be such—this is a separate issue—but it does not follow that they would play the central verifying role in evaluation. For some components in the valuational base—as we saw in the case of the central necessary conditions—verification consisted in showing that they were basic to the whole mass of diverse valuations of people in that age. (It does not matter how these latter are construed, that is, whether they are simply purposes and desires or feelings of pleasure or immediate prizings, or broad ideals.) In other cases, for example in dealing with needs, people's values in many senses entered as evidence into the account by which the existence of needs, their unsuppressible character and distorting effect if given no expression were established. This tied need assertions in a complex way to values, but not in such a way that every need or drive had to be prejudged as good without a further estimation of its role in human life.

The valuational base thus may be compared in ethical theory to the established body of systematic knowledge in a scientific field. It is a stable core, neither intuitively self-evident nor reducible to a definite set of particular experiences, but capable of refinement and extension both by theoretical analysis and additional experience.

N.B

A systematic treatment of ethical ideas and evaluation processes has not been attempted in this book, and must be postponed to a subsequent work. Accordingly, a fuller analysis of verification in ethics is beyond our present scope. The dominant emphasis has been on tracing the inroads of growing knowledge and possible knowledge in establishing more definite ethical answers. Now this very process of diminishing indeterminacy admits of interpretation through different models. A teleological philosophy will see it in terms of the magnetic power of an absolute good, and probably take the valuational base to be the inherent nature of man in its local manifestations. A mechanistic philosophy will see it as the convergence of causes to produce a unified effect, and a cultural relativism will add that it represents the growth of a single world culture stabilizing the goals of the scientifically oriented western world as a basis of judgment. (Regarding this as ultimately an arbitrary form would rest on a relativistic interpretation of knowledge itself.) A naturalistic philosophy would see the lessening of indeterminacy as a consequence of the growth of knowledge in the "genuine" sense of man's rational powers becoming more effective in the determination of his outlook. And so on. For the most part, we have not had occasion to go far into the different basic philosophical outlooks, and have concentrated on the effects of knowledge however this is itself interpreted philosophically. And it is in this same fashion that we may look in conclusion at the issue of residual indeterminacy.

§ Residual Indeterminacy

SOME INDETERMINACY WILL ALWAYS REMAIN. We have seen some of the ways in which the growing awareness of the biological, psychological, social and historical structuring of human life serves to lessen indeterminacy in ethical judgment. And we have seen that many avenues are open for further scientific investigation which may in time reduce the indeterminate zone even more. But, of course, indeterminacy has not been completely eliminated, nor can it ever be. Much of the judgment issues in generalizations which may not suffice for completeness in the particular situation. Although many of the differences among men may come to be seen in terms of permissible individual or cultural variation, others may involve a choice between interests in which deep values are lost whatever the direction taken. Sacrifice, tragedy in a multitude of forms, incalculable moral tension, not only have existed but will continue to exist in the lives of men. The diminution of indeterminacy does not mean, then, a moral calculating machine but at best at present, threads to follow, moorings to hold to, as one struggles with moral judgment. The hypothesis offered at the outset was that the trend of knowledge in the various fields had shifted, so that the center of gravity no longer lay in the fact of indeterminacy but in the structural lines pointing to increasing determinacy.

THE PROBLEM OF ORIENTATION TO RESIDUAL INDETERMINACY. Ultimate attitudes towards indeterminacy have always played an important part in ethical theory. For some, the indeterminacy could be well hidden in the confidence that all was shaped by a kindly divinity. By others it was felt as the intransigence of matter yielding despair, or the insensitivity of an alien or hostile cosmos. To some it brought the sense of novelty or the unexpected in the universe, the wide-open anything-can-happen feeling. Some felt it as the guarantee of irrationality, others as the supremacy of their self-assertion. Those found a most bitter taste in ethical relativity to whom it came with an ultimate it-does-not-really-matter sense; they felt as if all value had been drained from the cosmos.

Such qualities require careful and extensive evaluation. They have to be located and described, not only in general outline but in their specific cultural and historical form. Their causal roots have to be probed, and their consequences in human outlook and action estimated. This is the orientational aspect of every large-scale philosophy. In the light of the complex history of ethical relativity as a doctrine we cannot here undertake to explore it systematically.[17] But we can at least see what kind of attitude is appropriate towards residual indeterminacy on the thesis advanced in this book.

The characteristic feature of residual indeterminacy is that it is *residual*. That is, it is not tied to the core of ethical judgment, nor does it constitute its "essence"; it is only something left over. At best, if the solution is carried out to a number of decimal points, what is left over can be ignored. But even where it cannot be ignored it is not the answer but a shortcoming in the answer. In some cases, at

---

17. Some suggestions about the orientation effect were offered in Chapter I. See above, pp. 27-28.

worst, it ties application to incalculable hazards. But its quality remains a something-to-be-cut-down as solutions and applications become more refined.

One of the consequences of such an attitude is that residual indeterminacy loses its unity. In a thousand different ways, in different types of situations, it may be tackled. Sometimes it is cut down by increased knowledge. Sometimes its presence adds pressure to working out shared or common values. Sometimes it leaves no choice but taking a chance and bearing what may come. It may be that the precise quality of the attitude tends to come from the particular field, not from the element of indeterminacy as a central feature. Or if a general attitude is required, perhaps the one most called for on the approach we have sketched is an orientation of creativity.

ON BEING CREATIVE. To be creative means to approach a situation with the full sense of being an agent, a being capable of thought and action, who can bring to consciousness the manifold features of the situation, envisage objective alternative possibilities, and direct action to actualize his values. Such an attitude is not proposed here as an arbitrary preference. It would itself require full-length justification as against proposed alternatives, along the lines suggested by the successive contributions of the various perspectives. That is, psychological grounds could be offered for the creative attitude as against the purely passive-receptive or inactive-contemplative or indifferent come-what-may outlook, in each case by analysis of their modes of functioning and relation to the internal economy and dynamic development of the individual. Similarly, cultural standardizations of these alternative types would admit of functional evaluation. And socio-historical analysis would see these outlooks in terms of their perennial relationships and problem-solving contributions. We may offer the hypothesis that a full consideration of the creative attitude would find it the orientation congruent with the needs and core values that emerged in our study of the various perspectives.

In such an inquiry the fuller meaning of the creative attitude would emerge. Thus it would have to be distinguished from the attitude of arrogance that feels self-consciousness to be itself a mastery of the world, forgetful of its roots in nature. And it would have to be distinguished from the sense of blind spontaneity of a kind of naked individual self, as if its decisions issued from nothing but its isolated self, reflected nothing beyond, expressed no relations to the world of men.[18] As attitude it would have to embody the full awareness of the nature of the self, with its biological and psychological grounding, its social and cultural kinship and its full historical scope. So far from feeling itself as a lone self, it would see its position on the historical frontier of a developing world.

We are, in short, looking to the full scientific perspective which embraces the lessons of the human sciences from biology to history, and applies them to the contemporary life of society and the individual, to help us fashion a general outlook on our world and ourselves. This does not mean that "Science gives us values." Science does not create values, only men create values. Science does not give us virtues, but men grow virtues. Science does not give us goals, but men use their knowledge to broaden and refine and increasingly to achieve their human aims. And they use their growing knowledge of themselves to work out what their aims are and to distinguish increasingly the spurious from the genuine. A full scientific understanding thus molds their way of looking at the world. They see themselves at every point as active creators out of the past and into the future.

---

18. Cf. above, p. 154 f.

Index

Aberle, D. F., 226n.
Absolutism, ch. II; conflict with relativism, 29; indeterminacy in, 31-33; and moral law, 49-50
Abundance, effect in ethical theory, 320 ff.
Acton, Lord, 254
Adler, Alfred, 172
Adorno, T. W., 181n.
Aggression, 127 ff., 136, 139 f., 163, 171 f.
Alcoholism, 191n., 221
Alexander, the Great, 172, 271
Allport, Gordon W., 169, 172 f., 177n.
Amorality, psychological evaluation of, 187-192
Anarchism, 229
Anthropology, 38; contributions to diminishing ethical indeterminacy, ch. VII
Aquinas, Thomas, 42
Arbitrariness, sense of, 16-17, 27 f.; in ultimate premises, 72
Aristophanes, 20, 236
Aristotle, 21, 22, 42, 61, 62, 65, 133, 138, 189, 236, 262, 311, 330
Arrow, Kenneth J., 63n.
Asch, Solomon E., 168n., 177n., 193n., 296n.
Aspirations, see Ideals, Goals, Needs
Austin, J. L., 83n.
Authoritarian personality, 181 ff.
Authority, and conscience, 39; in moral law, 43-45; in conception of reason, 59
Ayer, A. J., 26, 83n.

Bacon, Francis, 262

Bali, 206, 224
Barker, R. G., 145
Bateson, Gregory, 206, 224
Beale, Howard K., 289n.
Beard, Charles, 264
Beebe-Center, J. G., 135n.
Behaviorism, 25, 38, 171
Benedict, Ruth, 26, 205n., 206n., 208, 213 f., 220 f., 223
Bentham, Jeremy, 24, 33 f., 62, 65 f., 132, 175
Bergson, Henri, 44
Berle, A. A., 322n.
Betsileo, 241n.
Bidney, David, 212n.
Biological, naturalization of moral phenomena, 121 ff.; models, 118; ethics, 120; criteria applied to cultural domain, 220 ff.
Biology, contributions to diminishing ethical indeterminacy, ch. V; contributions summarized, 120 f.
Black, Max, 99n.
Boas, Franz, 99n., 218n.
Boas, George, 268n.
Boulding, Kenneth E., 63
Bowlby, J., 173n.
Breaking-point, as criterion, 184 f.
Bridgman, P. W., 98n.
Brotherhood of man, 311-318
Burnham, James, 251n., 252n., 254
Butler, Joseph, 41, 65, 208

Caesar, 271
Cannon, W. B. 118
Cantril, Hadley, 177n., 179n.
Capitalism, 280, 322
Categorical imperative, 43

Causality, 79-82, 230-237
Cerf, Walter, 71
Change, 22 f.
Chartism, 276
Chastity, 204
Child, Irvin L., 146n.
Childe, V. Gordon, 263
Christie, Richard, 181n.
Cleckley, Hervey, 190n., 191n.
Cohen, A. K., 226n.
Cohen, Morris R., 99n.
Commitment, 71, 79 f.
Common-human morality, need for, 18; minimal survival type, 159 f.; bases for, Part III
Communism, 229, 280
Conflict, 29, 72, 88, 141
Conscience, 37-41; in the psychopathic personality, 190 ff.
Consciousness, of values, 162 ff.
Consistency, in ethics, 56; pragmatic, 57 f.
Contingent factors, role in ethics, 303 f.
Cooperation, role in survival, 117
Corporal punishment, 220
Creative attitude, 338 f.
Creative power, 268 f.
Cultural relativity, 26, 205-215; meaning, 205; critique of theses, 209 ff.; application to science, 282 ff.
Culture patterns, *see* Benedict
Custom, 19 f.

Darwin, Charles, 19, 38, 116, 117n., 139
Davis, A. K., 226n.
Death, attitudes to, 130 f.; death instinct, 127 ff.
Decadence, 267
Definition, 33, 76 ff.
Dembo, Tamara, 145, 198n.
Democracy, 229, 245, 325-328
Democratizing morality, 328
Dependence relations, 231 ff.
Determinism, 270 ff.; *see also* Causality, Dependence relations
Dewey, John, 96, 152 f., 258 f., 262
Dignity, 18, 209, 212 f., 327
Disagreement, ultimate, ch. III

Discriminations, 317
Disinterestedness, 64-67
Divorce, 238
Dostoyevsky, Fyodor, 2, 52
Doukhobors, 125
Drives, evaluation of biological, 137-143
Duncker, Karl, 135n., 193 f., 197 f., 205n., 218, 295n.
Duns Scotus, 32
Duties, 24, 42, 44

Ecclesiastes, 22
Economics, 62 f., 231 ff., 236 f., 240, 244, 263 f., 321 ff.
Edel, Abraham, 83n., 98n., 107n., 133n., 212n., 231n., 239n., 283n., 286n., 320n., 332n., 334n.
Educational theory, 25 f., 174, 194 f., 208, 328
Efficiency, 62 f.
Egoism, 23 f., 64-66, 313 f.; psychological evaluation of, 175-179
Emotive theory, 26 f., 34 f., 84 ff.; critique of, 86 f.
Ends, *see* Goals
English history, examples from, 272 ff.
Epictetus, 189
Equality, 18, 245, 273 f.
Erikson, Erik, H., 186n.
Eskimo, 210
Essences, search for moral, 203-205
Ethical invariance thesis, 194; evaluation of, 197 f.
Ethical relativity, ch. I, 109 ff.; conflict with ethical absolutism, 29; major arguments considered, ch. III; indefinite indeterminacy in, 93 f.; and functionalism in biology, 151 ff.; impact of theory of needs on, 173 ff.; implications of evidence concerning amorality, 191 f.; Duncker's formulation of, 193; avenues of socio-cultural inquiry into, 214 f.
Ethical theory, 122 f.; relation to historical context, 331 f.; its

evaluation in terms of valuational base, 332 f.; *see also* Absolutism, Ethical relativity

Ethnocentrism, 181 f.

Euripides, 20

Evil, 23, 124 f., 131 f., 135 f., 139, 144, 146, 184, 218, 307

Evolution, 116 f., 119, 139, 152 ff.

Experience, 18, 52 f.

Fact, relation to value, 74 f.; finding answers and making decisions, 106 ff.

Falk, W. D., 83n.

Fascist glorification of war, 309

Feigl, Herbert, 167n.

Fenichel, Otto, 129 f., 134, 180, 182n., 184 f., 190n., 191n.

Feuer, Lewis, 332n.

Field instability and complexity, 97

Flugel, J. C., 186

Fosdick, H. E., 261n.

Frank, L. K., 174n.

Frank, Philipp, 30n., 33n.

Franklin, Muriel, 177n.

Freedom, 18, 284

Frenkel-Brunswik, E., 181n.

Freud, Anna, 163n., 177n.

Freud, Sigmund, 38, 52 f., 128 f., 136, 163n., 171 f., 199

Fromm, Erich, 39, 135, 178, 200

Frustration, 58, 144-148, 220

Fuller, Lon L., 104n.

Functional autonomy, 138, 169

Functional theory, in biology and ethics, 148-155; sociology and ethics, 230 ff.; and ethical theory, 331 f.

Gandhi, on non-violence, 43 f.

Gearing, Fred, 240n.

Generality, in concept of reason, 67 f.

Gestalt theory, and ethics, 193-196

Goals, 25, 52, 70 f., 88 f., 216, 247 f., 300 f.

God, 32, 90, 312

Goldfrank, Esther S., 232n.

Goldstein, Kurt, 142, 169, 330n.

Good, 23 f., 29, 34, 256; apparent and real, 65; working conception of, 78 f.; Stevenson's model for, 85; life as, 124-132; drive expression as, 141 ff.; and evolutionary order, 153 f.; how discerned in historical trends, 248

Greek, historians, 19 f.; dramatists, 20, 236; philosophers, 21

Green, T. H., 175 f., 255

Group orientation, in biological perspective, 156 f.

Growth, 179

Guilt-feeling, 37, 38, 40

Gurvitz, Milton, 190n.

Haller, William, 274n.

Hallowell, A. I., 177n.

Happiness, 185 f., 223 f.; in psychopathic personality, 191

Hare, R. M., 83n., 96n.

Harmony, in concept of rationality, 60 ff.

Hart, H. L. A., 83n.

Hartmann, Heinz, 53n., 177n.

Hartmann, Nicolai, 32n.

Health, 222; as ideal, 143 f.; mental, 184 f., 220 f.

Hedonism, 24, 183; *see also* Pleasure, Utilitarianism

Hegel, G. W. F., 51, 90, 207n., 255, 271

Heisenberg, principle, 102

Hempel, Carl G., 77n.

Henderson, D. K., 127, 190n.

Herodotus, 19 f., 22, 26

Herskovits, Melville, J., 209

Historians' value criteria, 266 f.

History, contributions to diminishing ethical indeterminacy, ch. VIII; goal-patterns in, 247 f.; over-determination of ethical judgment in, 277; knowledge of, 287

Hitler, Adolf, 70, 197

Hobbes, Thomas, 16, 24, 100, 238, 252

Hobhouse, L. T., 257 f.

Höffding, Harold, 33n.

Hofstadter, Richard, 116n., 213
Hogg, Quintin, 279n.
Holmes, O. W., 116
Homeostasis, 118
Homosexuality, 180
Hopi, 240n.
Horney, Karen, 185n.
Horton, Donald, 221
Hull, Clark, 171
Hume, David, 51, 59
Humiliation, 220 f.
Humility, 163
Huxley, Julian, 17, 39, 119n., 139n., 141, 150, 153
Huxley, T. H., 152-154

Ideas, Plato's conception of, 51
Ideals, 18; generated by societal tasks, 228 f.; democratic, 235; how analyzed in power theory, 251, 253 f.; concept of moral ideal, 256; material bases of, 261 f.; Spartan, 174, 279; see also Brotherhood of man, Democracy, Justice
Immortality, 131
Imperatives, moral judgments as, 83
Impoverishment of personality, 185
Improvement, desire for, 179
Impulses, see Drives
Indeterminacy, in ethical theories, 28-35; nature and sources of, ch. IV; in history of science, 101 ff.; in legal theory, 104 ff.; diminished by human sciences, Part II; summary of modes of diminishing, 293 ff.; residual, 336 ff.
India, 223
Individualism, 23 f., 81 f., 116, 143, 157, 237, 257, 313 ff., 324; see also Egoism, Power
Inductive verification, 73
Industrialization, 223, 303, 319
Insight, 163, 222, 265
Instincts, as basis for ethics, 117; death instinct, 127 ff.; Darwin on, 139; see also Psychology
Intuitionism, 31 f.

Invariant societal tasks, 225-230
Invariant values, see Values
Irrationality, cult of, 16 f.
Israel, 240

Jahoda, Marie, 181n.
James, William, 140
Japan, 220
Jealousy, 163
Jefferson, 328
Job, 90
Justice, 18, 21, 205, 321 f.

Kadish, M. R., 96n.
Kant, Immanuel, 38, 41, 43-45, 56 f., 65, 67, 94, 189, 255, 311
Kaplan, Abraham, 252n.
Kardiner, Abram, 241n.
Karpman, Ben, 190n.
Kaufman, Felix, 243n.
Keith, Arthur, 117n.
Kennan, George F., 316
Killing, 126
Kinsey, A. C., 180
Klineberg, Otto, 118n.
Kluckhohn, Clyde, 78n., 216, 226n., 228n., 299n.
Knowledge, as ideal, 262; as historical effort, 265; relativity of, 282-289; historical, 287 f.
Köhler, Wolfgang, 195
Kroeber, A. L., 78n., 230n.
Kropotkin, Peter, 117, 158
Kwakiutl, 206

Lamont, Corliss, 131
Laski, Harold, 258n.
Lasswell, Harold D., 252n.
Laws, absolute, 32 f.; moral, 41-50
Lee, A. M. and E. B., 218n.
Lee, Dorothy, 170, 177n., 206n.
Leeper, R. W., 35n.
Legal theory, model from, 104 ff.
Lepley, Ray, 243n.
Levellers, 273 f.
Levinson, D. J., 181n.
Levy, M. J., 226n.
Lewin, Kurt, 145, 183
Lewis, C. I., 57
Lewis, Helen B., 176n.

Leys, Wayne R., 96n.
Liberalism, 143, 229, 237, 257 ff., 272
Liberty, 257-259, 324, 326
Life, value of, 124-132; paths of, 330
Lindner, R. M., 190n.
Linguistic issues, 26 f., 82 ff.
Linton, Ralph, 240n., 241n.
Lippincott, B. E., 258n.
Lippitt, Ronald, 183
Lochner v. *New York*, 116
Locke, John, 25 f.
Logical issues, in types of moral laws, 42 ff.; in concept of reason, 51 ff.; in status of ultimate premises, 72 f.; in relation of fact and value, 74 f.; in theory of definition, 76 f.; in emotive theory, 86 f.; in comparison of scientific and legal models, 106 ff.; in psychoanalytic procedure, 164 ff.; in theory of needs, 167 f.
Love, 61, 65, 163, 178
Lowie, R. H., 242 f.
Loyalty, 61
Luther, Martin, 80 f.; Luther-phenomenon, 81

MacBeath, A., 217, 231n., 256
Macdonald, Margaret, 83n.
McDougall, William, 118n.
McKeon, Richard, 325n., 326
Machiavelli, Niccolo, 24 f., 238
Mandelbaum, Maurice, 285n.
Manners, Robert A., 240n.
Mannheim, Karl, 285
Marshall, James, 252n.
Martin, C. E., 180n.
Marx, Karl, 52 f., 264 f., 271
Marxian theory, 25, 316
Maughs, Sidney, 190n.
Material control, as historical goal, 259-266
Maturity, 179 f.
Maximization, in concept of rationality, 62 f.
Mead, Margaret, 39n., 60n., 164n., 206, 223n., 224, 227n.
Mean, 61 f.

Means, 52, 62 f., 70 f., 126, 216, 302 f.
Means, G. C., 322n.
Mechanism, in psychology, 25, 192 f.
Mechanisms of defense, 163
Menninger, Karl, 127
Mental health, 184 f., 220 f.
Merton, Robert K., 219n., 230n.
Mesquakie, 240n.
Metaphysics, and objectivity of knowledge, 286
Mill, John Stuart, 38, 50, 66, 133, 176, 257, 326
Montagu, Ashley, 119n., 173n.
Moore, G. E., 32n.
Moral blindness, 189
Moral issues, reality of, 29
Moral law, 41-50
Morality, a human product, 19 ff., 121 f.; rise of, 122 f.; part of culture, 202 ff.; as historical goal, 254-257; democratization of, 328
Morris, Charles W., 330
Mosca, Gaetano, 25
Mullahy, Patrick, 163n., 172n.
Murdock, G. P., 226n.
Murphy, Gardner, 146 f., 223n.

Nadel, S. F., 231n.
Napoleon, 271
National interest, 316
Natural, 21, 183
Naturalization of moral phenomena, 121 ff.
Nazi, 16 f., 44, 48
Necessary conditions, in valuational base, 302 f.
Needs, psychological, 167-173; emotional needs, 223; in valuational base, 299 f.; as social principle, 323 f.
Nehru, Jawaharlal, 281n.
Neurosis, 86, 127, 163, 190, 199
New Guinea, 206
Nietzsche, F. W., 140, 172, 251
Normality, 183
Norms, 101-103, 122 f., 152 ff.; *see also* Value
du Noüy, Lecomte, 174

Obligation, 85; relation of *ought* to *is*, 74 f.; sense of, 122
Overstreet, H. A., 179n.

Pain, 58, 132-137
Pareto, Vilfredo, 25, 52
Parker, John, 279n.
Parsons, Talcott, 219n.
Pavlov, I. P., 25, 147
Peace, 160, 308 f.
Pericles, 268
Perry, Charner M., 71, 73
Perry, Ralph Barton, 61
Persians, 20
Personal base, 329 f.
Personality, anti-democratic, 181 ff.; maturity, 179 ff.; psychopathic, 189 ff.
Phase rules, 46-48
Phenomenological (phenomenal) description, 42 f., 193-196, 205
Piers, Gerhart, 39n.
Piety, 204, 216
Planck, Max, 102n.
Plato, 21, 38, 51, 58-60, 101 ff., 133, 137 f., 188 f., 192, 234, 252, 262
Pleasure, as good, 23 f., 34; evaluation of, 132-137; types of, 135; pleasure principle, 138
Pointer-readings, 40
Pomeroy, W. B., 180n.
Power, struggle for, 24 f., 250 ff.; will to, 172; critique of power conceptions, 252
Pratt, J. B., 64 f.
Predatory ethics, 116
Premises, ultimate, 72 f.
Problem-situations, general concept of, 96; historical structuring, 270, 280 ff.
Productive power, 235 f., 265, 319
Progress, 22, 186
Psychoanalysis, 86, 127-130, 136, 163-166, 172, 180, 184 f.
Psychology, 25, 38 f., 51, 127 ff., 134, 144 ff.; contributions to diminishing ethical indeterminacy, ch. VI; newer lessons for ethics, 161 f.; limitations in treatment of ethics, 199 f.; application of psychological criteria to cultural domain, 220 ff.
Pueblos, 232n.
Purity, 43, 204

Rashdall, Hastings, 94 f.
Rationalization, 52 f.
Reality, of moral values, 29; of pleasures, 133; reality principle, 138; real and spurious values, 164 ff.
Reason, 18, 50-69; components in ethics, 55-69
Regression, 145 f.
Regret, 57 f.
Relativity, in physics, distinguished from ethics, 30n.; *see* Cultural relativity, Ethical relativity
Relevance, problem of 269; concepts in philosophy of history, 270 ff.
Religion, 23 f., 138, 233 f., 237, 261; Freud on, 199
Repression, 142 f., 184
Respect for life, 48
Rhodes, Cecil, 172
Right, 32, 77 f., 86 ff.
Rivers, W. H. R., 228n.
Robespierre, 51
Roheim, Géza, 141
Roosevelt, Eleanor, 281n.
Royce, Josiah, 61
De Ruggiero, G., 258n.
Russell, Bertrand, 70 f., 251

Samoa, 164
Sanford, R. N., 181n.
Santayana, George, 61
Scarcity, 253; role in ethical theory, 320
Schlick, Moritz, 176
Schopenhauer, Arthur, 125
Schweitzer, Albert, 125, 312
Science, 36, 51; and ethical relativity, 15-17; and ultimate disagreement, 87 ff.; as model for coping with indeterminacy, 101 ff.; contributions to diminishing ethical indeterminacy, Part II; attempts to relativize, 210 f., 282-289; and value neutrality, 242 ff.; role in stabilizing a

common-human ethic, Part III

Self, 80 f., 176 f.; self-assertiveness, 154; selfishness, see Egoism

Semantic, see Linguistic issues

Sextus Empiricus, 21, 26

Shame, 38, 40

Sherif, Muzafer, 177n.

Sidgwick, Henry, 66

Simon, Herbert A., 96n.

Simpson, G. Gaylord, 117n.

Singer, Milton B., 39n.

Situational meanings, 194

Slavery, 21, 311

Smith, Adam, 34, 41, 65, 95

Social optimum, 63

Social science, and value neutrality, 242 ff.

Socialism, 229, 237, 279, 322

Sociology, contributions to diminishing ethical indeterminacy, ch. VII; sociology of knowledge, 285

Socrates, 38

Sophists, 21

Sorokin, P., 263n.

Spencer, Herbert, 116, 119n., 124, 134

Spengler, Oswald, 207

Spinoza, 332n.

Spitz, René A.,. 173n.

Spranger, Edward, 330

Stephen, Leslie, 118, 149

Stevenson, Charles L., 27, 83n., 84-86, 99n.

Stirner, Max, 176

Struggle for existence, 116

Subjectivity, 30, 34

Sublimation, 184 f.

Success, 204

Suicide, 126 ff., 220

Sumner, W. G., 26, 209, 212 f.

Survival of fittest, 116

Sutton, F. X., 226n.

Tanala, 241n.

Tawney, R. H., 23, 237

Tax, Sol., 212n., 240n.

Tension, 128 ff., 138, 141 f., 170; see also Drives

Thompson, Clara, 163n.

Thompson, Laura, 240n.

Thucydides, 20, 268n.

de Tocqueville, Alexis, 328

Tolerance, 26, 209, 212 f.

Tolman, E. C., 171

Toulmin, S. E., 70

Trans-cultural evaluation, 215-242

Trobriand islanders, 315

Truth, 285

Truth telling, 43, 47 f., 94 f.

Ultimate commitment, 71, 79 f.

Ultimate disagreement, ch. III

Ultimate premises, 72 f.

De Unamuno, Miguel, 131n.

Uniqueness, 30, 49

United Nations agencies, 324 f., 327

Unity, in conception of reason, 59; in valuational base, 306 f.

Universal rules in morals, 45 ff.

Usury, 21, 149

Utilitarianism, 33 f., 62, 132, 276

Valuational base, theory of, ch. IX; meaning of, 296 f.

Value, relation to fact, 74 f.; origins, 122; of life, 124-132; conscious representation of, 162 f.; and need, 170; neutrality in social sciences, 242-246

Value acquisition, phenomenology, of, 80

Value assumptions, in Sumner's relativity thesis, 212 f.; in Benedict's analysis of culture patterns, 213 f.

Values, real and spurious, 164 ff.; in cultural relativity thesis, 208 f.; search for invariant, 215 ff.; reorientation in, 240

Variety, in ethics, 29; in patterning of conscience, 38; in values, 82; in cultural forms, 205, 208

Veblen, Thorstein, 236

Vercors, 45, 48n.

Virtues, 24, 181, 203, 236 f., 240, 321

Vivas, Eliseo, 154 f.

Vogt, Evon Z., 240n.

Wallace, H. A., 281n.

War, moral equivalent of, 140
Waterhouse, Ian K., 146n.
Watson, John B., 25, 171
Weber, Max, 237, 243n.
Wegrocki, Henry J., 183n.
Westermarck, Edward, 34n.
White, Ralph K., 183
Whorf, B. L., 84
Will, to live, 125; to power, 172
Williams, Donald C., 71

Williams, Elgin, 214
Williams, Robin M., 219n.
Witchcraft, 221
Wittfogel, Karl A., 232n.
Wright, M. Erik, 145
Wrong, 32, 42, 47, 77, 84, 88

Zimmerman, C. C., 263n.
Zuckerman, S., 122n.

FREE PRESS PAPERBACKS

A Series of Paperbound Books in the Social and Natural Sciences, Philosophy, and the Humanities

These books, chosen for their intellectual importance and editorial excellence, are printed on good quality book paper, from the large and readable type of the cloth-bound edition, and are Smyth-sewn for enduring use. Free Press Paperbacks conform in every significant way to the high editorial and production standards maintained in the higher-priced, case-bound books published by The Free Press of Glencoe.

Andrews, Wayne	*Architecture, Ambition, and Americans*	90067
Aron, Raymond	*German Sociology*	90105
Bettelheim, Bruno	*Truants From Life*	90345
Cohen, Morris Raphael	*Reason and Nature*	90609
Coser, Lewis	*The Functions of Social Conflict*	90681
Durkheim, Emile	*The Division of Labor in Society*	90785
Durkheim, Emile	*The Rules of Sociological Method*	90850
Edel, Abraham	*Ethical Judgment*	90890
Eisenstadt, S. N.	*From Generation to Generation*	90938
Evans-Pritchard, E. E.	*Social Anthropology and Other Essays*	90987
Friedmann, Georges	*The Anatomy of Work*	91082
Friedmann, Georges	*Industrial Society*	91090
Geertz, Clifford	*The Religion of Java*	91146
Goode, William J.	*Religion Among the Primitives*	91242
Gouldner, Alvin W.	*Patterns of Industrial Bureaucracy*	91274
Hayek, F. A.	*The Counter-Revolution of Science*	91436
Henry, Andrew F., and James F. Short, Jr.	*Suicide and Homicide*	91442
Janowitz, Morris	*The Professional Soldier*	91618
Katz, Elihu, and Paul F. Lazarsfeld	*Personal Influence*	91715
Lerner, David, and Lucille W. Pevsner	*The Passing of Traditional Society*	91859
Maximoff, G. P.	*The Political Philosophy of Bakunin*	90121
Meyerson, Martin, and Edward C. Banfield	*Politics, Planning and the Public Interest*	92123
Neumann, Franz	*The Democratic and the Authoritarian State*	92291
Park, Robert Ezra	*Race and Culture*	92379
Parsons, Talcott	*Essays in Sociological Theory*	92403
Parsons, Talcott	*The Social System*	92419
Radcliffe-Brown, A. R.	*The Andaman Islanders*	92558
Reiss, Ira L.	*Premarital Sexual Standards in America*	92620
Riesman, David	*Individualism Reconsidered:* UNABRIDGED EDITION	92650
Rosenberg, Bernard, and David Manning White	*Mass Culture: The Popular Arts in America*	92708
Simmel, Georg	*Conflict* AND *The Web of Group Affiliations*	92884
Simmel, Georg	*The Sociology of Georg Simmel*	92892
Sorokin, Pitirim A.	*Social and Cultural Mobility*	93028
Wagner, Philip	*The Human Use of the Earth*	93357
Weber, Max	*The Theory of Social and Economic Organization*	93493

Many of these books are available in their original cloth bindings.
A complete catalogue of all Free Press titles will be sent on request